Farming in a Changing Climate

The Sustainability and the Environment series provides a comprehensive, independent, and critical evaluation of environmental and sustainability issues affecting Canada and the world today. Other volumes in the series are:

SUSTAINABILITY
AND THE
ENVIRONMENT

*Edited by Ellen Wall, Barry Smit,
and Johanna Wandel*

Farming in a Changing Climate:
Agricultural Adaptation in Canada

UBCPress · Vancouver · Toronto

15 14 13 12 11 10 09 08 07   5 4 3 2 1

Printed in Canada on acid-free paper that is 100% post-consumer recycled, processed chlorine-free, and printed with vegetable-based, low-VOC inks.

Library and Archives Canada Cataloguing in Publication

　　Farming in a changing climate: agricultural adaptation in Canada / edited by Ellen Wall, Barry Smit, and Johanna Wandel.

(Sustainability and the environment)
Includes bibliographical references and index.
ISBN 978-0-7748-1393-8 (bound); ISBN 978-0-7748-1394-5 (pbk.)

　　1. Crops and climate – Canada. 2. Agricultural ecology – Canada. 3. Agriculture and state – Canada. I. Wall, Shirley Ellen II. Smit, Barry III. Wandel, Johanna, 1971-IV. Series.

S600.7.C54F374 2007　　　　333.76'0971　　　　C2007-905170-7

Canadä

UBC Press gratefully acknowledges the financial support for our publishing program of the Government of Canada through the Book Publishing Industry Development Program (BPIDP), and of the Canada Council for the Arts, and the British Columbia Arts Council.

This book has been published with the help of a grant from the Canadian Federation for the Humanities and Social Sciences, through the Aid to Scholarly Publications Programme, using funds provided by the Social Sciences and Humanities Research Council of Canada.

UBC Press
The University of British Columbia
2029 West Mall
Vancouver, BC V6T 1Z2
604-822-5959 / Fax: 604-822-6083
www.ubcpress.ca

Contents

Figures and Tables

Tables

Abbreviations

BSE	Bovine spongiform encephalopathy
C-CIARN	Canadian Climate Impacts and Adaptation Research Network
CERES	Crop Environment Resource Synthesis
CGCM	Canadian Global Coupled Model with versions (e.g., CGCM1 and CGCM2)
CGCMI-A	Canadian Global Coupled Model that includes the effects of aerosols
CSIROMk2	Commonwealth Scientific and Industrial Research Organisation Mark 2
EPIC	Erosion Productivity Impact Calculator
FAO	Food and Agriculture Organization
FAQ	Financière agricole du Québec
GCM	Global Climate Model or Global Circulation Model
GDD	Growing degree days
GDP	Gross domestic product
GFDL	Geophysical Fluid Dynamics Laboratory
GHG	Greenhouse gas
GISS	Goddard Institute for Space Studies
HadCM	Hadley Centre Coupled Model; also HadCM2, HadCM3
IIASA	International Institute for Applied Systems Analysis
IPCC	Intergovernmental Panel on Climate Change
LSRS	Land Suitability Rating System
MPI	Max Planck Institute Circulation Model
NGO	Non-governmental organization
P	Precipitation
PCM	Parallel Climate Model
PE or PET	Potential evapotranspiration
PFRA	Prairie Farm Rehabilitation Administration
RCM	Regional Climate Model

SDSM	Statistical Downscaling Model
SRES	Special Report on Emission Scenarios
SSRB	South Saskatchewan River Basin
UKMO	United Kingdom Meteorological Office
UPA	Union des Producteurs Agricoles du Québec

Preface

Extreme weather events have taken a substantial toll on human livelihoods and lives around the globe, and have often detrimentally affected food production and security. In 2005, persistent droughts in several African countries severely limited food supplies, flooding in Bangladesh routinely disrupted agriculture, heat spells in Australia caused crop and livestock losses, and North American hurricanes such as Katrina led to significant crop losses and blocked grain transportation systems. With climate change, the expectation is that temperatures will rise, moisture conditions will change, and many extreme climate events will become more common. Given the effects of recent extreme weather, questions arise about the capacity of agri-food systems to handle changed climate and weather in the future. Such capacity may be found in individuals and families, local communities, regional authorities, business and corporations, and/or national governments. All have a part to play in preparing for challenges – both risks and opportunities – from future climatic and weather conditions.

In Canada, indications are that climate change is already having an effect on farming, thereby increasing the need for research and programs to assist adaptive decision making. Several groups in the Canadian agri-food sector seek relevant and timely information. One is industry-related, including producers and agribusiness interests who view climate and weather risks as one of several factors to be considered in operating strategies affecting farm production practices and financial management. Another is made up of policy makers charged with the task of developing programs and legislation that can enhance the agri-food sector's ability to manage climate risks and take advantage of opportunities. A third group is the research community, which seeks to improve the understanding of the implications of climate change for the agri-food sector and to provide a sound basis for making decisions about adaptive strategies.

To date, information about climate change impacts and adaptation has for the most part been fragmented, in terms of both the issues focused on (for instance, temperature, crops, adaptive strategies) and location, with some areas in Canada receiving more attention than others. Notwithstanding the interconnections among research focus and region, there has been little opportunity to compare different research perspectives, analytical methods, and results relevant to climate change adaptation in Canada's agri-food sector. To address these knowledge gaps, C-CIARN Agriculture (Canadian Climate Impacts and Adaptation Research Network for Agriculture) sponsored a workshop in Edmonton, Alberta, on 17 February 2005. At this event, leading Canadian scholars and researchers presented and critiqued their latest research findings on climate change adaptation and agriculture. A panel of agricultural producers and policy makers also offered their comments on the utility of the results presented.

This book is a direct result of that workshop; it provides a review and summary of the current state of knowledge about climate change impacts and adaptation for the Canadian agri-food sector. It also identifies research gaps and issues that need to be addressed if policy and programs for agricultural adaptation to climate change are to be timely and effective. Material presented in this volume provides a comprehensive look at the issues and research relevant to anyone interested in climate change and agricultural adaptation.

This book is divided into five parts. The first is devoted to introductory issues, including information on different research approaches for studying climate change adaptation and agriculture. Three dominant research perspectives are noted and form the basis for organizing the material presented in the next three parts. Part 2 presents research findings that consider adaptation using scenario-based impact approaches. Part 3 illustrates research that places climatic stresses in the context of the many forces that influence agriculture. The research described in Part 4 is also contextual, focusing on farm-level sensitivities and adaptation processes. The conclusion in Part 5 includes comments from representatives of the agriculture industry and policy makers.

Acknowledgments

We are grateful to many individuals and institutions whose contributions have resulted in the publication of this book. Thank you to the chapter authors, who took the time and had the patience to work through the details of preparing their work for publication. We appreciate comments and suggestions from two anonymous reviewers, and help from the staff at UBC Press in guiding us through the book production process. Several students and staff at the University of Guelph assisted in different ways in preparing drafts for submission. Special thanks to Marie Puddister and Monica Hadarits.

We are indebted to financial support from the Social Sciences and Humanities Research Council of Canada (SSHRC), including the Aid to Research Workshops and Conferences in Canada, Aid to Scholarly Publications Programme, Standard Research Grants Programme, and Multi-Collaborative Research Initiatives Programme. We also appreciate funding from the Government of Canada's Climate Change Impacts and Adaptation Program and the Canadian Climate Impacts and Adaptation Research Network, as it has been essential not only for preparing this volume but also for supporting much of the research and discussion reflected in the chapters. We are also thankful for research support from the Ontario Ministry of Agriculture, Food and Rural Affairs (OMAFRA) and the Canada Research Chairs Program. Several of the chapters in this book include acknowledgments from their authors for particular contributions to the research in the chapter.

We also wish to express our gratitude to the many agricultural producers who often gave us the benefit of their time and experience in our attempt to understand the complexities of agricultural adaptation and climate change in Canada. We trust that our interpretations will be useful and contribute in some way to the future prosperity of the Canadian agri-food sector.

Part 1:
Research Approaches to Climate Change Adaptation

Chapters 1 to 4 outline the issues underlying climate change adaptation research in agriculture and offer ways to differentiate among approaches to understanding those issues. The classification employed is an organizing principle for presenting research findings relevant to climate change adaptation in Canadian agriculture. The approaches are not mutually exclusive, but the various studies tend to rely mainly on one of the three orientations.

One approach, sometimes referred to as "top-down," starts with climate change scenarios and examines adaptations in light of the potential impacts of the specified climate changes. We label this approach "impact-based." Another perspective emphasizes the various factors (physical, economic, social, institutional) influencing adaptation decision making, and assesses climate effects in the context of those forces. This orientation is labelled "context-based." A related perspective, sometimes called "bottom-up," focuses on the process of adaptation, particularly the role of farmers and other stakeholders, and examines their capacity for managing risks and adapting to changes. We label this approach "process-based." These categories are used to distinguish the main foci of the various studies, although researchers often employ features from more than one perspective.

1

Introduction

Ellen Wall, Barry Smit, and Johanna Wandel

Climate, Weather, and Canadian Agriculture

The history of agriculture involves a continuous series of adaptations to a wide range of factors. Environmental conditions related to soil, water, terrain, and climate impose constraints and provide opportunities for agricultural production, while technological developments lead to modifications in the structure and processes of farming operations. Likewise, market factors related to input costs and prices have a dramatic effect on what commodities are produced and how and where production takes place. Public policies and programs are also major elements influencing the agri-food sector.

None of these factors remains constant and their effects are interdependent. Their changes over time represent stimuli that affect the success of farming activities and that prompt adjustments to altered circumstances. Variable climatic and weather conditions are fundamental to crop and livestock production in the agri-food sector (Kandlikar and Risbey 2000). Patterns of temperature, moisture, and weather conditions greatly influence plant and animal performance, inputs, management practices, yields, and economic returns. As grain producer Brett Meinert (2003) put it, "we harvest water and sunshine." Adaptation to climate and weather risks is therefore implicit in the ongoing development of the agri-food sector. With climate change, growing conditions and climate-related risks and opportunities are expected to change, and may already be changing (Rosenzweig et al. 2000).

For many, the phrase "climate change" is associated with, and related to, the analysis, policy, and action concerning increased greenhouse gas emissions and subsequent global warming (for example, Bhatti et al. 2006). Since 1997, representatives from Canadian government departments, industrial sectors, and the research community have devoted human and financial resources to developing policies and programs related to the greenhouse gas commitments under the Kyoto Protocol. Their efforts have been aimed at reducing or mitigating greenhouse gas emissions to address one dimension of the climate change issue, namely, the need to slow down or stabilize climate change.

Concurrent with mitigation is another climate change response or objective: to develop and promote adaptation strategies that will reduce the adverse effects of climate change itself and moderate the risks while capturing opportunities associated with changing climatic and weather conditions (Burton et al. 2002; Smit et al. 2000b). The United Nations Framework Convention on Climate Change (UNFCCC) includes obligations related to greenhouse gas emissions reduction and carbon sequestration (mitigation), and also to adaptation: "to formulate, implement ... and regularly update ... programs containing measures ... to facilitate adequate adaptation to climate change" (Article 4.1). The UNFCCC also requires signatories to "cooperate in preparing for adaptation to the impacts of climate change, develop and elaborate appropriate and integrated plans for coastal zone management, water resources and agriculture" (Article 4.1) (UNFCCC 1992).

The *Climate Change Plan for Canada* (November 2002) is mostly about emissions reductions or mitigation, but it also includes a commitment to "develop and research approaches to adaptation planning and tool development" and to "develop increased awareness of the impacts of climate change and the need to address them in the future through adaptation."

Adaptation in the agri-food sector involves practices, programs, and policies that reduce vulnerabilities to climate and realize opportunities (Bryant et al. 2000). Several players are implicated in adaptation initiatives, including producers, agribusiness, and government agencies (Adger and Kelly 1999; Smit and Skinner 2002). In Canada, the Standing Senate Committee on Agriculture and Forestry notes that "a general goal of government policies should be to encourage the adoption of opportunities to adapt to climate change" and recommends incorporating "climate change [adaptation] into existing policies and programs ... under the category of 'no regret' policies" (SSCAF 2003, 68-69).

Research to date has identified substantial risks and some opportunities from climate change for the agri-food sector across Canada. Depictions of climate change often focus on increases in temperature, suggesting that the main effect will be gradual warming. For some Canadian regions, this could be beneficial if it results in production opportunities from an extended growing season and increases in available heat units. More heat and a longer season should allow for increased flexibility in timing of operations and choice of crops or varieties, particularly on northern margins. For instance, Québec and Ontario producers have been able to expand grain production with plant cultivar development, and expect corn and soybean production to extend to more northern regions. Although the opportunity to extend agricultural production northward is appealing and often assumed possible, soil quality, moisture availability, and other constraints may impede such developments (Kandlikar and Risbey 2000).

The focus on temperature changes reflected in the term "global warming" tends to mask other significant alterations in conditions associated with climate change, particularly changes in the frequency, magnitude, and extent of climatic extremes such as droughts and floods (Smit and Pilifosova 2003). Climate is naturally variable and agricultural systems have evolved to cope with modest variations in conditions, but they are susceptible to extremes. It appears that, in some regions, the extremes associated with climate change are pushing coping capacity to its limits. A cash crop farmer from southwestern Ontario notes:

> Weather is getting to be more sporadic and unpredictable each year. In the last three years with any luck, you probably got the best crops your farm has ever seen and if you are the other people your crops have been disastrous. Weather has been extreme and it seems to be tough to get that gentle rain that you need. Lots of areas are getting almost nothing and other areas are getting what we call hundred year storms. Something is wrong when you get three, hundred years storms in five years. (K. Oke, pers. comm.)

A fruit grower from British Columbia makes a similar observation:

> Many farmers, including myself, see hail as a big problem for us. Twenty or 30 years ago, hail events occurred maybe once every eight or ten years. My farm has been hailed seven times in the last 10 years. That is fairly typical. It is quite substantial ... All I see is that weather events are more intense, and the frequency of these weather events is increasing. This is coming at a time, unfortunately ... where our crop insurance premiums have just doubled. We have a big problem with this because with increased weather events that affect our crops and our ability to grow good quality crops, we want affordable crop insurance. (Patton 2003)

Patton's comment illustrates the fact that producers make direct links between weather conditions and farm financial management. Despite the important economic consequences from climate change, it is often characterized as primarily an environmental issue and impacts are defined in terms of temperature zones, production conditions, growing season conditions, and/or yields. As noted by Patton (2003) and Oke (pers. comm.), however, climatic and weather conditions pose risks for the financial viability of individual farming enterprises, regional agricultural sectors, and rural communities, depending on agricultural activity. Also affected are agribusiness firms that supply inputs, process outputs, and provide services, and the institutions that fund support programs related to agricultural production.

Climate change adaptation in Canadian agriculture is a topic that has a broad application and far-reaching consequences.

Adaptation Options for Managing Climate and Weather Risks

Climate change adaptations are adjustments in management strategies to actual or expected climatic conditions or their effects, in order to reduce risks or realize opportunities. They can take many forms, can occur at different scales, and can be undertaken by different agents (producers, agribusiness, industry organizations, and governments). Adaptations are not necessarily discrete technical measures, but are often modifications to farm practices and public policies with respect to multiple (climatic and nonclimatic) stimuli and conditions.

Some climate change researchers include adaptation choices in their assessment of "impacts," recognizing that the severity of climate change risk depends on the responses of producers and other agri-food sector players. Implicit in such models is the so-called "smart farmer" assumption, namely, that producers have knowledge of climatic conditions in advance and perfect adaptations are instantaneously adopted (Adams et al. 2003; Easterling et al. 1992a; Smit 1991). For example, McKenney and colleagues (1992) assume farmers' adaptation responses for the MINK (Missouri, Iowa, Nebraska, and Kansas) region in United States. Using the EPIC (Erosion Productivity Impact Calculator) Model, they create a future baseline for crop productivity in the year 2030 that reflects changes based on technological advances. These new technologies include several crop-breeding improvements that lead to higher yields, more efficient chemical conversions, and earlier leaf development. Also assumed are projected improvements in pest control and harvesting techniques (reducing losses). In some cases, additional adjustments are used, such as crop substitution and additions, alterations in planting dates, and more efficient irrigation. With these assumed conditions, outcomes for three of four major crops (soybean, wheat, and sorghum) suggest enhanced performance in 2030 under climate change, while corn yields are projected to decline. Without adaptations and adjustments, all yields are projected to decline (McKenney et al. 1992).

Similarly, Easterling and colleagues (1992b) study the effectiveness of adaptation and adjustments at the farm level by running impact models (EPIC) under future climate scenarios. They compare effects on costs and revenues for cases with and those without alterations in farm production practices. For the most part, the assumed adjustments and adaptations to projected climatic conditions offset the otherwise negative impacts, even when increased input costs are incorporated in the analysis. Easterling and colleagues (1997) also estimate the effects that shelterbelts will have for grain production under altered climatic conditions in the Great Plains region of US.

Using the EPIC Model with projected climate features (precipitation, temperature, and wind speed), the authors find the "shelterbelt effect" to be positive, especially for regions with severe precipitation deficiency and highly increased wind speeds.

Research conducted for the Canadian Prairies, based on modelling for an average climate change year, concludes that adopting management strategies (such as changing to a different crop and earlier seeding) makes a substantially positive difference with few exceptions (Cloutis et al. 2001). Antle and colleagues (2004) draw similar conclusions using data from Saskatchewan in their impact assessments for Great Plains agriculture. Nagy (2001) reports on the possible consequences for energy use in farming systems in the region when two different adaptation options are introduced into the model, namely, diversifying crops and altering nitrogen use. The PCEM (Prairie Crop Energy Module) was modified to include increased acreages of two new crops, chickpeas and dry beans. Results indicate that introducing these crops into rotation may lead to reduced nitrogen and energy use (Nagy 2001).

These modelling-based analyses suggest that adaptation to climate change can play a significant role in moderating impacts on the agricultural sector. By assuming one or several types of adaptations, researchers have demonstrated that climate change presents not only challenges but also opportunities for farmers. Adaptation options are an integral part of climate risk management and can be examined in terms of what is possible (typologies of adaptation strategies) and what is done (how producers adapt).

Typologies of Adaptation Strategies

Early work in identifying types of practices to deal with climatic and weather conditions designated short- and long-term measures to counteract the impact of drought in the Great Plains region of North America (Rosenberg 1981). The latter includes minimum tillage, snow management, irrigation scheduling, microclimate modification through windbreaks, diversification of crops, improved production practices (for example, crop rotation, alternative planting methods, timing of fertilization), and crop breeding. Also using short- and long-term categorization, but in more recent documentation, Kurukulasuriya and Rosenthal (2003) generate a "matrix of adaptations" for agriculture applicable on a global scale. Included in short-term options are a variety of farm-level responses such as crop insurance, diversification, adjustments to the timing of farm operations, changes in cropping intensity, alterations in livestock management practices, conservation tillage, and efficient water use. Long-term strategies tend to focus on industry and state action. For example, Kurukulasuriya and Rosenthal (2003) list the following as long-term strategies: technological developments, agricultural pricing and market reforms, trade promotion, enhanced extension services, weather forecasting mechanisms, and a general strengthening of institutional and decision-making structures.

Without reference to duration, Smit and Skinner (2002) offer a comprehensive account of possible adaptation options for Canadian agriculture. The authors organize their findings according to four possible types. The first two (technological developments and government programs and insurance) apply mainly to options at the industry and state level, while the last two (farm production practices and farm financial management) focus on farm-level management.

Diverse adaptation options for producers will ultimately depend on what is feasible and realistic (André and Bryant 2001; Bryant et al. 2000). Many of the choices available are also closely linked to practices already in place for maintaining economic and environmental sustainability. Acknowledging the connections between "sustainable agriculture" practices and climate change adaptation helps to streamline policy and programs for both issues (Wall and Smit 2005). The concept of combining adaptation strategies with established decision-making processes that address other goals for the sector is similar to the notion of "mainstreaming" climate change adaptation with development in the international context (e.g., Huq et al. 2003).

Adaptation from Producers' Perspectives

As the research focus moves from crop yield impact models to the management of climate risks in the farm business, more attention has been focused on producers' perspectives and experience. In some studies, producers are asked to identify changes to production practices that result in benefits when faced with recent climate and weather risks. Some Ontario producers have noted that, in their opinion, climatic and weather conditions have changed noticeably in the past five years. Among other actions, their responses to such conditions include growing different crops and/or crop varieties, altering tile drainage, employing conservation tillage, changing the timing of planting, and installing irrigation systems when water availability is adequate (C-CIARN Agriculture 2002).

Also in Ontario, but with reference to soybean production only, Smithers and Blay-Palmer (2001) identify farm production practices that producers have adopted, thereby reducing risks from specific climatic stresses. Strategies include planting new or improved crop varieties that stand up under adverse climatic and weather conditions, adapting crop rotations, and altering the timing of planting.

A number of tactics have been employed to manage climate and weather-related stress at the community level. In southern Ontario, for example, producers joined forces with local water managers and developed a framework for participatory water management committees to ensure both the fair sharing principle and the maintenance of flows for ecosystem services (Shortt et al. 2004). These "irrigation advisory committees" were formed to deal with recent decreases in streamflows and increased water takings for irrigation. A number of similar committees have been formed in neighbouring areas where drought conditions prevail (Shortt et al. 2004).

Processing tomato producers in southwestern Ontario adopt measures to increase their production efficiency in light of drought stress. These include improved irrigation systems adapted from Australia, where conditions are much drier than in Ontario, to reduce the impact of extended droughts (AAFC 2003). In the 2002 season, one of the driest years in history, Ontario tomato growers with the new system had their second-highest yield ever (AAFC 2003).

Other researchers investigate specific adaptation options to explore their implications for practice and policy. For instance, Bradshaw and colleagues (2004) identify several constraints to crop diversification, including new and additional costs associated with technology required for different production systems, the pressure to specialize to meet economies of scale, better returns from diversifying "off the farm" through pluriactivity, and biophysical and locational limitations related to soil type and distance from markets. Despite such barriers, crop diversification in some regions of Canada (such as the Prairies) has taken place when viewed at the regional scale. At the individual farm level, however, there is little evidence that producers employ diversification tactics when faced with financial and production risks. Similar results have been noted for European agriculture (EU Commission 2001). Policy and programs encouraging producers to diversify their farm operations need to take into account other factors (such as the established trend towards specialization) that can work against such actions.

Climate Change Adaptation and Agriculture Research: Different Approaches

Research into climate change adaptation began to emerge as a research focus distinct from climate change mitigation in the late 1980s and early 1990s (Smit 1993). The main concerns shaping early inquiry included identifying what adaptations would likely be employed to limit or offset impacts from climate change. Adaptation issues were also important for policy applications based on information about possible strategies and how to evaluate their merit (Smit et al. 2000b). .

Research on climate change adaptation and Canadian agriculture varies according to characteristics of agriculture, such as commodity type, production system, or region, and according to the scale of analysis, from plant or plot through farm to region or globe. Adaptations can be distinguished by the type of climatic condition considered, such as temperature, moisture, or extreme events. Studies can also be grouped by their focus on physical conditions, biological variables, or social and economic processes, and whether they are primarily theoretical, modelling, or empirical analyses. Inquiry also differs according to research orientation, with some efforts contributing mainly to scientific knowledge while others are explicitly applied or policy-oriented.

This book recognizes the great variety in analyses, and differentiates them broadly according to the types of questions they are addressing and their starting points for examining adaptation. One type of inquiry focuses on four questions: "What are the impacts of expected climate change?" "How serious is climate change?" "What adaptations could possibly address the estimated impacts?" and "How much of the impacts can be moderated or offset by particular adaptations?" Research addressing these questions tends to start with selected climate change scenarios, models the impacts of these conditions on particular aspects of agriculture (commonly local agroclimatic conditions and yields), and then includes hypothetical or assumed adaptations in the modelling to estimate the "residual impacts." This "top-down" research approach is labelled here as *impact-based approach*.

A complementary perspective addresses the questions "What are the conditions that affect producers, and how do they deal with them?" "What facilitates or constrains adaptation in practice?" and "What is the context in which adaptation in agriculture occurs?" This perspective identifies the climate and non-climate forces that influence decisions in agriculture and documents the role of multiple forces (climate, economic, social, etc.) and multiple scales that define the context within which adaptation occurs. Such studies are considered here under the *context-based approach*.

A third type of research addresses the questions "How do the processes of adaptation work?" "Who makes the decisions?" "To what conditions are adaptations undertaken?" "What conditions influence the types of adaptations employed or not employed?" "What is the prospect of these adaptations being viable under future conditions?" and "What can be done to facilitate adaptation in practice?" This approach begins with understanding how the farming system experiences exposure to hazards (such as climatic conditions), how it is vulnerable, and how adaptation decisions are made. These insights into adaptation processes can be applied to assess the capacity for adaptations to future climate change. This "bottom-up" research approach, which focuses on decision making at the local scale, is referred to here as the *process-based approach*.

These three broad perspectives are used to structure this book, although many studies do not fit cleanly into a single category. Several studies primarily follow one of the approaches but incorporate elements from another. Nonetheless, these categories do provide distinct types of information about what climate change means for Canadian agriculture, how climatic conditions affect agriculture, and the prospects for adaptation in the sector.

Outline of the Contents of This Book

The focus of this book is on adaptation to climate change in Canadian agriculture. This introductory chapter has established that climate change is an important issue for the Canadian agri-food sector and has introduced three approaches for understanding agricultural adaptation to climate change. The other chapters in Part 1 provide more in-depth illustrations of these perspectives.

Chapter 2 describes the history and development of the impact-based approach, including its strengths and limitations. A survey of research employing this approach provides a comprehensive look at what impact-based research can accomplish. The conclusion to Chapter 2 points out the value of addressing underlying causes of vulnerability. This topic is examined more fully in Chapter 3, where the authors outline the context-based approach and review research results in terms of risk and the role of government programs and policy. Chapter 4 concludes Part 1 by focusing on the process-based approach. Although elements of this perspective are well established in methodologies frequently employed in social sciences, some of the practices have only recently been employed in climate change adaptation and agriculture research. Consequently, Chapter 4 provides more detail on the theoretical and conceptual elements of the process-based approach and less on results from such studies in Canadian agriculture.

Part 2 contains three examples of studies broadly following the impact-based approach for climate change and Canadian agriculture. Chapter 5 (by Samuel Gameda and colleagues) uses climate change scenarios for impact-based assessments relying on different agroclimatic indices and their implications for specific types of crop production in Atlantic and central Canada. In Chapter 6, David Sauchyn considers Prairie agriculture and presents historical data on climate impacts as well as current and future effects. Chapter 7 is based on a major impact study of the Okanagan Basin in British Columbia. In that chapter, Denise Neilsen and colleagues demonstrate how scenarios of future climate can be used to estimate impacts and engage stakeholders in discussions about adaptive responses to potential impacts.

Context-based studies are featured in Part 3. In Chapter 8, Ben Bradshaw examines how climate risks are treated in light of other risks for primary agriculture and offers insights into expected adaptations, given the historical trajectory of Canadian agriculture. Henry Venema picks up on this theme in Chapter 9 and provides specific details from Prairie agriculture to make the point that adaptive capacity for climate change impacts is embedded in a host of social and economic factors. In Chapter 10, Harry Diaz and David Gauthier focus on the importance of adaptive capacity at the institutional level, particularly contributions from government programs. They use findings from research in the South Saskatchewan River Basin to illustrate their points. Christopher Bryant and colleagues (Chapter 11) conclude Part 3 with their assessment of climate change adaptation in Québec agriculture. They offer a review of the research on the topic that leads directly to current Québec research focusing on the use of crop insurance for managing climate and weather risks and the policy implications related to that adaptation strategy.

Part 4 includes examples of studies about climate change adaptation and Canadian agriculture that follow the process-based approach. In Chapter 12, Suzanne Belliveau and colleagues reveal how British Columbia apple and grape production is subject to a host of conditions and factors that affect strategic decision making for all risks, including weather and climatic conditions. This finding is also relevant to research from Ontario presented in Chapter 13. In that chapter, Susanna Reid and colleagues document producers' experiences with climate and weather risks in a specific region of the province. Their analysis demonstrates both the wide variety of current climate adaptation strategies and their integration with existing farm practices and management issues. Cynthia Neudoerffer and David Waltner-Toews (Chapter 14) also use rural community examples for their case study of the process of building capacity and resilience in a Manitoba farming region. This examination of residents' responses to past soil erosion and flooding problems points out several features that need to be in place if and when climatic and weather conditions bring renewed stress to those areas. In Chapter 15, Robert McLeman employs a historical case study to show how climate change may affect rural agricultural population patterns. He uses community experience in Oklahoma during the 1930s Dust Bowl to demonstrate the processes underlying rural families' ability to move from the affected regions. McLeman points out that such a strategy is a form of adaptation to adverse climate and weather impacts with possible applications for contemporary and future Canadian rural agricultural communities.

The concluding section, Part 5, consists of two chapters. Chapter 16 contains observations from agricultural producers and policy representatives who attended the workshop that gave rise to this book. Their insights on climate change adaptation research for the Canadian agri-food sector are reported as verbatim commentary and organized according to three issues: Variability/Uncertainty, Capacity, and Adaptation Processes. Chapter 17 provides highlights from the material covered in the previous sixteen chapters. This summary is offered as a comprehensive overview of the issues and gaps in our knowledge regarding climate change adaptation and Canadian agriculture.

2

Impact-Based Approach

Michael Brklacich, Barry Smit, Ellen Wall,
and Johanna Wandel

Human activities have been altering the chemical composition of the earth's atmosphere for several centuries (Ruddiman 2005). Increased levels of "greenhouse gases" (GHGs) such as carbon dioxide and methane are an undeniable marker of human-induced atmospheric change (Weaver 2004). GHG concentrations in the atmosphere have been increasing exponentially for about 200 years, and it is now apparent that human activities are contributing to climate changes on a planetary scale (Beade et al. 2001; Crutzen 2002).

The prospect of global climate change has sparked considerable interest in relationships between climate change and human activities in the scientific and policy communities. Given that climate is a major input to crop production, it is not surprising that agriculture has received considerable attention with respect to climate change (e.g., Parry 1990). Some of the earliest assessments of the impacts of climate change on human systems were undertaken for the agricultural sector (Adams et al. 1988; Arthur and Abizadeh 1988; Rozensweig 1985; Smit et al. 1988, 1989). Climate change and agricultural relationships featured prominently in the first three assessment reports issued by the Intergovernmental Panel on Climate Change (IPCC) (Gitay et al. 2001), and syntheses of the relationship between climate change and Canadian agriculture have been produced (Brklacich et al. 1997a; Bryant et al. 2000; and Natural Resources Canada 2002).

As noted in Chapter 1, an array of approaches and methods exist to examine the relationships between climate change and agriculture. Of particular interest in this chapter is the impact-based approach, which Dessai and Hulme (2004) refer to as the "standard approach" given its promotion by the IPCC (Carter et al. 1994) and its dominant use in assessments during the 1990s. The IPCC recommended selecting and applying climate change scenarios so that an assessment of biophysical and socio-economic impacts could be made relative to a standard set of future climatic conditions. After estimating yield impacts, researchers would then suggest possible adaptation strategies to limit the projected effects of climate change.

The purpose of this chapter is to describe and illustrate the main components of the impact-based approach and to provide an overview of climate change adaptation and agriculture research conducted from this perspective. The strengths and limitations of the approach are noted and opportunities for advancing our collective understanding of the relationships between climate change and agriculture are suggested.

Characteristics of the Climate Change Impact-Based Approach

A sequential assessment of climate change/agriculture relationships typifies impact-based approaches (Figure 2.1), with the usual departure point involving the specification of a macro-climate change scenario. Macro-climate change estimates are derived from the application of General Circulation Models (GCMs) under perturbed conditions, such as the commonly used doubling of atmospheric concentrations of greenhouse gases in carbon dioxide equivalents, or "$2 \times CO_2$." A variety of GCMs are available, including the Canadian General Circulation Model (CGCM), the Goddard Institute for Space Studies (GISS) model, the United Kingdom Meteorological Office (UKMO) model, and the Hadley Centre Coupled Model (HadCM). These models typically generate estimates of future climate norms, particularly temperatures for relatively large grid cells (for example, 3° latitude by 3° longitude). Currently there are many attempts to "downscale" GCM output and provide a more detailed set of climate change estimates.

Other methods have also been employed to specify future climates, including the use of spatial and historical analogues. For example, conditions in more southerly locations in the United States act as a surrogate for southwestern Ontario under increased mean temperatures. Researchers have also used conditions from the 1930s to portray a future that is characterized by more extreme weather events, especially drought, in the Canadian Prairies. In addition, some researchers include estimated changes in ground-level CO_2 as part of a future scenario, in recognition of the role that CO_2 enrichment can play in plant growth.

Outputs from these macro-climate change scenarios are inconsistent with the data required to conduct agricultural impact assessments from at least two perspectives, and must therefore be transformed. Many agricultural sectors comprise multiple activities and are often quite heterogeneous over short distances. Hence, the relatively coarse climate change scenarios must be downscaled to a regional level. In addition, the standard climatic properties, such as changes from current monthly averages for minimum and maximum temperatures, that are included in the macro-climate scenarios need to be converted into agroclimatic parameters that are more directly pertinent to crop growth and maturity (such as start and end dates for the frost-free season and growing degree days during the frost-free season). Overall, these transformations of relatively coarse macroclimate data provide

Figure 2.1

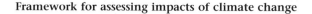

Framework for assessing impacts of climate change

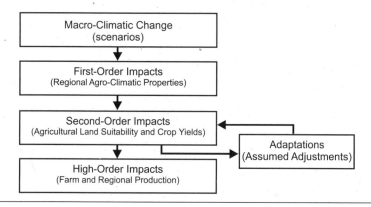

a basis for assessing the effects of human-induced climate change on regional agroclimatic properties as well as providing critical information for subsequent regional assessments of impacts on agricultural land suitability and crop yields.

Agricultural land suitability and crop yield represent two major categories of second-order impact assessments. Land suitability is the coarser of the two and provides estimates of the inherent or biological capacity of a particular location to support a major crop group, such as small grains, oilseeds, or tender fruits. The future climatic conditions are considered relative to soil, heat availability, and moisture to assess changes in land suitability for crop production. Crop yield assessments, on the other hand, are more specific and estimate the outputs (e.g., kg/ha) for a specific crop (e.g., winter wheat, canola, or apples), under the scenario's projected climate. In each case, these approaches to resource assessment employ either statistical or biophysical models that effectively compare available inputs, such as soil, heat, and moisture, with a crop's requirements for growth and maturation. The application of these models to climate change issues begins with a baseline assessment, which replicates current conditions, and then climate-sensitive parameters are adjusted to reflect a future climate. The difference between these model runs is then interpreted as the impacts stemming from the a priori specified change in climate. These approaches have been developed and refined considerably over the past twenty years. A recent study by the Food and Agriculture Organization (FAO) and the International Institute for Applied Systems Analysis (IIASA) (Fischer et al. 2001) includes an example of the current generation of second-order impact assessments that provide insights into the possible effects of a range of climate change scenarios on cereal yields.

These assessments of the impacts of climate change on land suitability and crop yields also provide a foundation for higher-order impact assessments. For instance, future-yield estimates are used to infer economic implications and possible crop shifts. The models are initially run under current or baseline conditions. Subsequently, climate-sensitive parameters included in the farm model (e.g., crop yields, irrigation water demands) are adjusted to reflect the climates specified under the macro-climatic scenarios in order to estimate the effects of climate change on profit levels (making assumptions about input costs and product prices). On a broader scale, higher-order impact assessments can employ regional production potential models to investigate the capacity of large regions to produce major crops under current conditions and then compare these estimates of baseline potential with production potential under an altered climate.

Overall, the impact-based approach consists of sequential or hierarchical stages that commence with the specification of a macro-climate change scenario and then trace the effects of this altered climate on basic inputs to agricultural production systems, land suitability and crop yields, and, finally, regional production potential or economic returns.

Key Attributes of the Impact-Based Approach

Several attributes characterize impact-based approaches for climate change and agricultural relationships. This perspective commonly considers climate change stressors on agricultural systems and assumes that human-induced climate change will proceed slowly and incrementally over several decades. Alternatively, climate change may be considered in a comparative static way, comparing average climates several decades apart. From an agricultural perspective, the approach tends to focus on a single spatial scale, either the farm-level or a regional scale, and considers socio-economic forces in rather simplistic terms; that is, farmers are assumed to adopt certain technologies and land-use practices, and have the managerial skills (and economic and other resources) required to implement particular adaptation strategies. Finally, this approach considers the vulnerability of farms, or the agricultural sector indirectly, as residual impacts; that is, vulnerability is defined as the negative impacts that adaptation cannot ameliorate. In this context, vulnerability is viewed as a residual value (vulnerability = impacts – adaptation).

The impact-based approach is well suited to exploring the physical impacts on aspects of an agricultural system, such as crop yields or regional production potential, of such climatic stressors as changes in average seasonal heat availability. The range and variety of studies employing this approach in Canada are evident in the next section of this chapter, which contains a comprehensive overview of research conducted from the impact-based perspective. Material is presented in terms of how climate change might affect three different phenomena: agroclimatic properties, types of production, and/or types of farming systems and regional economies.

Review of Canadian Climate Change and Agriculture Research Using the Impact-Based Approach

Impacts on Agroclimatic Properties

The agroclimatic properties assumed to be most important for the agri-food sector include growing and frost-free seasons, seasonal values for temperature, growing degree days, corn heat units, precipitation, and moisture deficits (Brklacich et al. 1997a). Most regions in Canada are expected to experience warmer conditions, longer frost-free seasons, and increased evapotranspiration rates:

> There is strong consensus that global climate change will result in longer and warmer frost free periods across Canada and thereby generally enhance thermal regimes for commercial agriculture. These changes in agroclimatic conditions are not expected to impact regions on an equal basis, with the longest extensions of the frost free season expected in Atlantic Canada. The extent to which these longer and warmer frost free seasons might benefit Canada, however, will in all likelihood be diminished by increases in seasonal moisture deficits across all regions and under all climate change scenarios. Hence it is crucial that all assessments of the implications of global climate change for Canadian agriculture take account of the possibility of both negative and positive impacts on agroclimatic properties. (Brklacich et al. 1997a, 233)

Changes in moisture conditions are expected concurrently with increased temperatures, particularly an increase in the frequency of extremes such as dry spells. Figure 2.2 illustrates that the return period for severe droughts in Western Canada is decreased with climate change. Under current climatic conditions, for example, a 30-day dry spell is expected to occur every 50 years. By 2070, dry spells of this length can be expected every 20 years. Similarly, the return period for 20-day dry spells might decrease from once every 35 years to once every 15 years.

In most of North America, the average temperature is expected to increase and the number of rain days to decrease under most scenarios, which means that there is an associated risk of increased frequency of long, dry spells (Gregory et al. 1997). Projections for specific Canadian regions indicate wide variation across the country. West coast estimates are for warmer and wetter conditions. Precipitation may increase in the winter months but decrease in the summer. Scenario projections for the Prairie region indicate

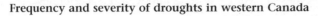

Figure 2.2

Frequency and severity of droughts in western Canada

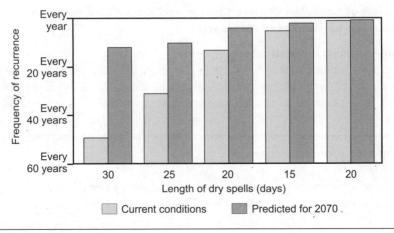

Source: After Hengeveld (2000).

a return to historic conditions, where persistent aridity was recorded for intervals of decades or longer (Sauchyn et al. 2003). More specifically, analysis of drought risks for the southern Saskatchewan area suggests that soil moisture conditions could become more variable, with the frequency of severe drought and drought conditions increasing dramatically (Williams et al. 1988).

Future conditions for Manitoba are similar to those for Saskatchewan, with suggestions that winters may be warmer and have less snow; summer temperatures are expected to rise; and precipitation, although reduced, may come more often in major events. It is likely that the growing season will be extended but may include more frequent extreme temperatures and severe hailstorms. Like Manitoba, Ontario and Québec are expected to have longer growing seasons (DesJarlais et al. 2004). Ontario will likely experience warmer winter and summer temperatures, increased extreme weather events (e.g., violent storms), and prolonged dry spells (Koshida and Avis 1998). Québec may experience an increase of precipitation in northern regions while levels are expected to remain constant or decrease in the southern part of the province (Koshida and Avis 1998).

Conditions in parts of the Atlantic region appear to be more difficult to project. It may be that conditions will be warmer and wetter in future years, extending the growing season and providing more heat units for crop production (Bootsma et al. 2001). Precipitation levels will likely vary widely, possibly leading to drought conditions in some areas while other areas might be too wet (Koshida and Avis 1998).

Although there is a great deal of uncertainty surrounding projections from climate change scenarios and their possible impacts on agroclimatic properties across Canada, various studies provide some relatively consistent results. One way to apply the expected changes in agroclimatic properties is to examine the potential consequences they could have for crop and livestock production.

Impacts on Crop and Livestock Production

A substantial body of research goes beyond agroclimatic effects to estimate impacts on crop and livestock production in light of climate change scenarios and associated elevated levels of CO_2. This approach focuses on climate change impacts on agricultural resources and biophysical yields, usually assuming no change in technologies and management practices.

Crop Yield and Production

De Jong and colleagues (1999) use the CGCM1 (Canadian Global Coupled Model 1) to determine changes in temperature and precipitation for a number of agricultural regions across Canada. They employ the Erosion Productivity Impact Calculator (EPIC) Model to predict crop yields under the estimated climate norms. Based on the expectation of warmer and slightly wetter climatic conditions, a one- to two-week advancement of planting dates for eastern and central Canada and approximately three weeks in the west is projected. Potential outcomes for yield show no significant change for barley, wheat, and canola, while corn nitrogen fertility needs in central Canada could increase. Soybean, potatoes, and winter wheat yields are projected to increase substantially.

For British Columbia, research into climate change impacts on agriculture in the semi-arid interior valley regions has focused on irrigated crop production. Based on projections from the CGCM1, Neilsen and colleagues (2001) estimate that production will be disadvantaged unless there are improvements in water supplies and irrigation efficiencies. They also consider potential impacts on fruit production, noting that the growing season could be extended by more than one month. Such a change could favour some apple and grape production, as would the reduction in potential harm from winter damage. Other challenges might arise from these factors however, such as harm to produce from extreme heat and from the persistence of pests able to survive a milder winter.

In the Prairie region, Williams and Wheaton (1998) examined potential climate change impacts for Saskatchewan based on climate models from the GISS and the IIASA. Biomass productivity and wind erosion potential were estimated using CA (climatic index of agricultural potential) and C (wind erosion climatic factor). Results indicate that the peak growth season would arrive earlier and that there would be more risk of wind erosion in midsummer. Although increased rainfall may offset any negative impacts, suggestions are that, with further warming, trends in biomass potential would be variable across time and provincial region (Williams and Wheaton 1998).

More recent analysis of the same region produces somewhat contradictory results for crop production (McGinn et al. 1999; Nyirfa and Harron 2001). Based on climate projections from the CGCM1 and crop yield estimates from EPIC, there appear to be opportunities for wheat production in an average year related to an advance in possible seeding and harvesting dates. A potential 50 percent increase in the number of growing days across the region is estimated for 2040-2069 (McGinn et al. 1999). These positive results also include effects from elevated high atmospheric CO_2. By contrast, Nyirfa and Harron (2001) used the CGCM1 in conjunction with the Land Suitability Rating System (LSRS) and concluded that constraints from moisture deficits and heat would offset the advantages of predicted precipitation increases during the same period.

Brklacich and Stewart (1995) incorporated information from the GISS, GFDL (Geophysical Fluid Dynamics Laboratory), and UKMO (United Kingdom Meteorological Office) models at double CO_2 concentrations for their analysis of climate change impacts for the Prairie region. They note that each model projects a different set of agroclimatic conditions and therefore varying impacts on wheat production, estimated with a CERES-wheat model (Crop Environment Resource Synthesis): "Temperature increases would lengthen the frost-free season and reduce the risk of frost damage, but the higher temperatures would hasten the crop maturation process and thereby suppress yields. Elevated CO_2 levels would improve water use efficiency (WUE) and provide more Carbon for photosynthesis, and thereby tend to offset the potential negative effects of shortened crop maturation periods" (Brklacich and Stewart 1995, 155).

In addition, the authors examine the effects of specific adaptive strategies (irrigation, winter wheat conversion, and earlier seeding) that producers are assumed to adopt. Irrigation appears to be the most effective (if feasible and sustainable) response for offsetting losses associated with the climate scenarios. Conversion to winter wheat would be beneficial in southern sites, given the possibility of more effective use of early spring moisture. Earlier seeding options, while being the easiest to implement, are the least likely to have widespread positive results because other factors (i.e., temperature and moisture stress) could suppress yields. Similar conclusions are presented in the work of Delcourt and van Kooten (1995), who employed a different circulation model, the CGCM2, and focused only on study areas in southwestern Saskatchewan (part of the Palliser's Triangle). Their analysis suggests substantial wheat yield loss and erosion of the farming economy under climate change.

For Ontario and Québec, projected changes in agroclimatic properties (based on the use of CGCM2) may have potential benefits for corn and sorghum produced in southern Québec, but these benefits are less likely for wheat and soybeans in this region (El Maayar et al. 1997; Singh et al. 1998). Similar conclusions are drawn for corn yields in regions of the midwestern United States (Southworth et al. 2000), where conditions are like those in southern Ontario. In this US case, predictions from HadCM2 (Hadley Centre Coupled Model 2) and Centre for European, Russian and Eurasian Studies (CERES) models are consistent and led researchers to conclude that corn yields in an average year would change significantly, with northern areas experiencing gains and southern regions losses. Strzepek and colleagues (1999) modelled water use and corn production using circulation models from the GISS, GFDL, and Max Planck Institute (MPI). Their analysis is based on output from WATBAL for water supply, WEAP (Water Evaluation and Planning System) for water demand forecasting, and CERES-Maize, SOYGRO, and CROPWAT for crop and irrigation modelling. They conclude that the current relative abundance of water in the region will likely be maintained up to the 2020s but find that progressively larger changes in the 2050s and beyond may compromise water availability for irrigation.

Climate models indicate that areas close to the Great Lakes Basin are expected to have a warmer, wetter climate (Andresen et al. 2000), while analysis using DAFOSYM (Dairy Forage System Model), CERES-Maize, and SOYGRO suggests possible northward extension of crop production and dramatic increases in yields for soybeans and maize. Results employing the HadCM2 and CGCM1 indicate that yields for some forages may also improve and fruit production in the area might benefit from extensions in growing season length and seasonal heat accumulation (Winkler et al. 2000). These results focus on climate normals, however, and do not include potential effects from inadequate fertility and/or new pest infestations, which have the potential to strongly affect production. Reliance on average temperature and precipitation rates tend to mask site-to-site and year-to-year variability in yield (Kling et al. 2003).

Such limitations were taken into account for a study on fruit production in the Great Lakes Basin that includes downscaling the CGCM1 and HadCM2 to finer spatial and temporal scales (Winkler et al. 2002). The authors incorporated relevant agroclimatic factors, such as the frequency and timing of threshold events (e.g., fall and spring freeze dates) and increased risks from pests, in the analysis. When such elements were included in estimations for codling moth development, it was not certain that climate change would bring to the area more amenable conditions for fruit production (Winkler et al. 2002). On the contrary, there is substantial evidence that Great Lakes regions may remain vulnerable to springtime cold injury and experience heavier pest infestations.

In the Atlantic region, Bootsma and colleagues (2001) used the CGCM1 and concluded that it was likely that crop heat units would increase substantially. They projected increases in yields for grain and soybean with little change indicated for barley. Also relevant to eastern Canada are findings from Bélanger and colleagues (2002), who employed the same climate models to project warmer winters, which may harm perennial forage crops by reducing the amount of protective snow cover and increasing the occurrence of above-freezing temperatures. At the same time, having warmer temperatures in the fall could reduce the cold-hardiness of perennial plants.

Additional Research on Possible Impacts on Crop Production
Scenarios and climate models are not the only tools available for estimating possible climate impacts on crops from future climate change. Basing their findings on the general expectation of increases in temperature and precipitation, some researchers conclude that climate change could have implications for plant disease and crop production in three ways: direct losses from diseased crops, challenges to plant disease management, and geographical distribution of plant diseases (Chakraborty et al. 2000). Similar factors are important for insect pests; climate change is expected to increase the migration, reproduction, feeding activity, and population dynamics of insects and mites, thereby leading to crop losses.

Coakley and colleagues (1999) note that there are serious issues regarding climate change and plant disease management. A review of key findings indicates that precipitation has more pronounced effects on plant disease than temperature, yet current GCMs cannot provide the necessary details on precipitation events. Also challenging is the difference in temporal and spatial scales for plant disease and climate models (Chakraborty et al. 2000). Climate change is expected to affect the incidence and severity of plant diseases in a number of ways, including the survival of pathogens, the rate of disease progress during a growing season, and the duration of the annual epidemic in relation to the host plant (Boland et al. 2003).

All crop production entails some degree of pest, disease, and nutrient management. Research has been conducted into how such practices might be and are affected by different climatic and weather conditions. For instance, Archambault and colleagues (2001) investigated changes in the efficacy of commonly used herbicides under increased temperature and CO_2 concentrations based on controlled experiments. They conclude that although there is a potential for herbicides to be less effective, the possible increase in crop yield may, in fact, offset any negative outcome. Ziska (2004) also addresses questions regarding weed persistence in changing climatic conditions and finds that invasive weed species show a strong growth response to recent and projected increases in atmospheric CO_2 but also a weakened efficacy of chemical control. Pattey and colleagues (2001) conclude that consideration of weather variables is advisable for effective nitrogen management in corn production, information that takes on more importance in light of potential change to climatic and weather conditions.

Crop production eventually results in foodstuffs for human consumption. Matters related to the quality of food products can also be a climate change issue. Research indicates that crops grown in elevated CO_2 levels may lack micronutrients essential for human health (Lawton 2002). This could lead to negative effects from the direct ingestion of plants deficient in trace minerals such as iron, zinc, chromium, and magnesium. It also has implications for human consumption of food created from animal products, if those animals are fed material that is deficient in micronutrients (Loladze 2002).

Scenario-based and other future-impact assessments for crop production constitute the bulk of research on climate change impact and adaptation to date. Results demonstrate wide variation in outcomes depending on the models employed and assumptions made. It is clear, however, that climate change poses serious risks and some opportunities for crop yields. The next subsection considers how forage and livestock production might be affected.

Forage and Livestock Production
The use of climate models and scenarios to project direct impacts on livestock is rare. Such models have been employed, however, to examine possible impacts on forage production (Adams et al. 2003) and grassland sustainability, both of which are important factors for livestock production. For instance, Baker and colleagues (1993) assessed potential effects on ecosystem processes and cattle production in US rangelands incorporating output from the GFDL, GISS, and UKMO into various ecosystem simulation models. Their analysis for the more northern regions projects a 10 percent decrease in soil organic matter with an increase in nitrogen available for plant uptake. For cattle grazing, this could have positive results related to using forage more and relying less on food supplements, especially in the spring months. The authors raise questions, however, about the long-term sustainability of such systems, given the loss of organic matter in the soil and increased variability in plant production (Baker et al. 1993).

Cohen et al. (2002) used the CGCM1 and a forage production model, GrassGro Decision Support System (DSS), to estimate the effects of projected climatic conditions on livestock production in three Saskatchewan regions. Their analysis includes different adaptation strategies related to choice of plants in pasture mixes. Results demonstrate strong variability across regions and plant type, but indicate that some grazing systems in Saskatchewan may benefit from climate change, if the adaptation options are viable and adopted.

Additional Research on Possible Impacts on Forage and Livestock Production
As noted for crop production, impact assessments have also been made using more general attributes of future climate change. For instance, after modifying levels of CO_2, nitrogen deposition, precipitation, and temperature in experimental plots, Zavaleta and colleagues (2003) concluded that changes to grassland diversity (and therefore grazing availability) may be rapid. The authors replicated plausible future conditions and concluded that while small increases in temperature have no obvious effect, additional CO_2 or nitrogen rapidly decrease species diversity. Their results also illustrate the additive nature of the effects of combined treatments. For example, plots that received both CO_2 and nitrogen exhibited twice the decrease in diversity compared with plots that received just one of the treatments. Indications that soil moisture can increase when the species diversity declines were also noted. Their work confirms the importance of considering many climate change factors simultaneously (Zavaleta et al. 2003).

Future climatic conditions also have direct implications for livestock production (Wolfe, n.d.). Increases in heat stress, for instance, could result in lower weight gains and milk production in cattle/cows, lower conception rates for all livestock types, and substantial losses in poultry production (Adams et al. 1998, cited in Kling et al. 2003). Furthermore, increases in extreme events (e.g., violent storms and flooding) might result in livestock losses (Kling et al. 2003), and high daytime temperatures can reduce total grazing time (Owensby et al. 1996). Water supplies for livestock can be negatively affected by changes in quantity and quality. In extreme drought conditions, the potential for water to become toxic from sulphur and *Cyanobacteria* (blue-green algae) creates serious problems for cattle production (PFRA 2003).

Charron and colleagues (2003) review the potential risks for livestock production resulting from the effects of climate change on animal diseases. Table 2.1 summarizes some of the possible disease outcomes for livestock related to altered climatic and weather conditions.

Charron and colleagues (2003) note that alterations in rainfall patterns and temperatures affect the chances of survival and enhancement of insect vectors (ticks, mosquitoes) and associated diseases previously considered exotic or rare (West Nile virus, leishmaniasis). Milder winters can reduce the prevalence of some problems, such as pneumonia in adult cattle. There are, however, greater chances of many more problems increasing as several diseases in young livestock (e.g., pneumonia and diarrhea) respond to rapid changes in temperature and moisture rather than to slowly increasing (or decreasing) averages. Milder winters can influence parasite survival in and on animals, adding to existing parasite loads. Livestock may be also be affected by contaminated runoff in watersheds where heavy rainfall (and/or flooding after drought) flush bacteria and parasites into water systems.

Table 2.1

Potential effects of climate change on vector- and non-vector–borne infectious diseases in animals

Disease agent	Animals at risk	Transmission	Effects of climate variability and change
Bacterial diseases			
Anthrax (*Bacillus anthracis*)	Domestic animals (especially herbivores)	Water-borne, food-borne, inhalation	Spores are highly resistant to altered conditions. There may be improved environmental conditions for dissemination and concentration of spores.
E. coli enteritis (*Escherichia coli*)	Livestock (calves, lambs, kids, pigs, foals)	Water-borne, food-borne	Stress due to environmental conditions precipitates disease. Flooding can increase distribution.
Leptospirosis (*Leptospirosis* spp.)	Cattle, horses, swine	Water-borne	There may be improved environmental conditions for proliferation of organism. Flooding can increase distribution.
Salmonellosis (*Salmonella* spp.)	Livestock	Water-borne, food-borne	Organism proliferates in warmer conditions. Flooding can increase distribution.
Tuberculosis (*Mycobacterium bovis*)	Livestock	Inhalation, food-borne	Hosts range might expand. Disease agent survives well in cold, damp conditions.
Yersiniosis (*Yersinia enterocolitica, Yersinia pseudotuberculosis*)	Sheep, pigs, goats	Water-borne, food-borne	Stress due to environmental conditions precipitates disease. Flooding can increase distribution.

▼ Table 2.1

Disease agent	Animals at risk	Transmission	Effects of climate variability and change
Viral diseases			
Influenza A (*Orthomyxovirus*)	Pigs, poultry, waterfowl, horses	Aerosol or direct	Potential increase in habitat, range, and abundance of reservoir hosts increases risk of interspecies transmission.
Bluetongue (*Orbivirus*)	Sheep, domestic deer	Not applicable	There may be altered geographic distribution of vector species. There may be enhanced vector competence, potential for creating new vector species, increased availability of breeding sites for vectors, increased passive airborne dispersal of vector.

Source: Modified from Charron et al. (2003).

Climate change scenarios suggest substantial challenges and some benefits for crop and livestock production in Canada. The potential impacts vary widely across regions and commodity type. Possible changes in specific aspects of farm production will have consequences for farming systems and their related regional economies. These are reviewed in the next subsection.

Impacts on Farming Systems and Regional Economies
Climate and weather impacts on crop and livestock production inevitably have consequences for the regional economies within which farming systems function. Most estimates of impacts on regional agricultural economies are based on temperature and moisture norms from climate change scenarios. Assessments of climate change traced through agroclimatic conditions and yields are aggregated and suggest that impacts on the North American agricultural economy may be minimal when compared with those expected for less developed nations (Wolfe, n.d.). The IPCC (2001, 56) reports with "high confidence" that for North America, "small to moderate climate change will not imperil food and fibre production," while cautioning that there will likely be wide variation in impacts within the continent. For the US, Reilly and colleagues (2001) used Hadley and Canadian models to determine the likelihood of extreme events (more hot days and fewer cold days; more heavy rain or longer droughts) and assessed results. Assuming that producers make necessary adaptations, they conclude that climate change in the US will have an overall positive effect on production, with a resulting drop in prices likely leading to challenges for producers (especially those who do not practise adaptation). Adams and colleagues (2003, 131), however, using the RCM (Regional Climate Model), note that positive predictions for climate change impacts are highly questionable given that "assessments based on finer scale climatological information consistently yield a less favourable assessment of the implications of climate change."

Reinsborough (2003) applied a Ricardian analysis for climate change scenarios for Canada, and incorporated projections from the CGCM1. This approach assumes that spatial associations between temperature and other climate norms on the one hand, and agricultural land values on the other, will apply under changed climate, reflecting autonomous adaptations in the agri-food sector. Her work built on US analysis by Mendelsohn and colleagues (1994) that concluded that there may be overall benefits to the agricultural economy with projected climate change. Findings from Mendelsohn and colleagues (1994) were based on the balancing of cropland impacts (which appear to result in 4-5 percent losses) with crop revenue impacts (which appear to result in 1 percent gains). By contrast, Reinsborough (2003) found that gross agricultural revenue under climate change scenarios could improve or decline by 6.4 percent. Such a large margin of error is exacerbated by the difficulties encountered when incorporating realistic adaptation costs (Reinsborough 2003).

Reinsborough's analysis follows other work on the impacts of climate change on Canadian agricultural economy in terms of costs and benefits. Cline (1992) and Kane and colleagues (1992) report negative impacts on the agri-food sector from climate change impacts, while Nordhaus and Boyer (2000) suggest a modest benefit. Reilly (1995) also projects net benefits for Canada as long as assumed adaptation options are pursued and CO_2 fertilization is incorporated in the scenario. Weber and Hauser (2003), using downscaled projections from the CGCM2, concur with Reilly et al. (2003). They base their conclusions on an improved agricultural gross domestic product (GDP) (from increases in land values) in all provinces, suggesting that Reinsborough is too pessimistic.

Much of the research relevant to impacts of climate change on Canadian farming systems and regional economies has been for Prairie regions, where climate change is predicted to have major impacts (Chiotti 1998). Because agriculture plays an important economic role in the area, stresses and opportunities for the sector are considered significant (Cloutis et al. 2001). Among the earliest assessments of the potential effects of climate change on Prairie agriculture is the work of Arthur and Abizadeh (1988). Their analysis relies on the GFDL and GISS circulation models for climate change and the Versatile Soil Moisture Budget (VSMB) for determining crop responses. Their analysis builds on earlier work by Williams and colleagues (1987, cited in Arthur and Abizadeh 1988), who conclude that substantial losses will ensue for the sector. Arthur and Abizadeh (1988) note that, as long as adjustments are made to take advantage of potential opportunities from these changed conditions, the outcomes could be positive (especially in northern areas). Schweger and Hooey (1991) used the GISS and IISA output to estimate effects on soil erosion, and conclude that there are serious concerns about escalating erosion and salinity in the Prairies connected to potential increases in moisture deficits.

Changes in growing conditions associated with future climatic and weather conditions would have direct effects on the viability of Ontario farming systems. Brklacich and Smit (1992) applied GISS for the climate model and a Cropping Budget System Model in their analysis. They note that extended frost-free seasons and more variable precipitation will likely take place and pose considerable risks for crop production in Ontario. Advantages from longer growing seasons may be offset by reductions in moisture levels, resulting in fluctuating farm income levels and reduced capacity for food production in the province. Brklacich and colleagues (1997b) emphasize producers' adaptation responses to determine possible impacts from altered climatic and weather conditions. The authors combine a number of climate change scenarios to produce a mid-range depiction of plausible climate changes for a specific region in Ontario. Producers from two types of farming systems (livestock and diversified) responded to the future scenarios in terms of how they might alter their farming systems to take advantage of conditions or lessen potential risks. Results suggest that adaptation options would be pursued, with livestock operators (whose farms tended to be larger than diversified operations) potentially adopting a wider range of actions than diversified farmers.

Kling and colleagues (2003) base their assessment of the effects of climate change on Ontario farming systems on two GCMs, the Parallel Climate Model (PCM) and HadCM3. They note that risks to producers would inevitably result from increasing year-to-year climate variability. This is congruent with the findings of Brklacich and Smit (1992), who suggested greater fluctuations in farm profits resulting from variability in precipitation and extended frost-free seasons. Kling and colleagues (2003) also indicate that small to medium-sized operations will be more disadvantaged in higher-risk circumstances.

Projections from climate change scenarios have been used to assess future possibilities for farming systems in specific agricultural regions of the Annapolis Valley in Nova Scotia (Mehlman 2003). Applying the CGCM1 and a Statistical Downscaling Model (SDSM), future conditions (including precipitation, frost-free days, hot days, and extremely hot days) were estimated for three future time periods. In general, spring months in the farming areas are expected to be warmer and drier, while summer, autumn, and winter months will likely be warmer and wetter. Indications are that extreme events coinciding with hurricane season will be more frequent during fall, and that there will be a substantial increase in days with above-freezing temperatures in winter. The potential increase in growing season length suggests positive outcomes for farming systems, but these may be offset by the negative effects of either too much or too little moisture and extreme events.

Researchers adopting the impact-based approach use climate scenarios to estimate future challenges and opportunities for the Canadian agri-food sector, whether at the level of farm production or for the broader agricultural economy. How these challenges and opportunities (as well as other factors not captured in the scenarios) will be met depends in large part on producer and institutional adaptive strategies and risk management. These topics are discussed in later chapters.

Beyond the Impact-Based Approach

Agricultural research and agri-food policy have evolved and matured considerably over the past twenty years, primarily in response to the growing awareness that national agricultural economies, including Canada's agri-food sector, are increasingly being influenced by a broader set of forces that originate from within and beyond agricultural sectors. These changes have resulted in several calls for climate/agricultural research to build upon this foundation. Four key themes that would strengthen climate/agricultural research can be identified:

(1) *Broaden the context.* There is a need to move beyond the focus on climate change alone. Agri-food systems are subjected to multiple stressors that are constantly shifting. Thus, to accurately assess the role of climate, climate change assessments must be situated within the complex and dynamic environment in which agriculture operates. Future assessments need to relate not only to future climate scenarios but also to other climatic, environmental, social, economic, and policy conditions, including development pathways (i.e., changes in international trade policies, food preferences, etc.) that are expected to shape the future of agriculture.

(2) *Consideration of multiple spatial and temporal scales.* There is also a need to move beyond the consideration of single spatial and temporal scales in isolation. It is now widely recognized that assessments of climate change on complex agri-food systems must differentiate between fast- and slow-moving parameters and recognize the role of local and broader forces and responses (Clark 1985). For example, climate change models are designed to estimate conditions over several decades, whereas technological advances and economic globalization have been compressing the time associated with major shifts in agricultural production. Research into climate change and agriculture needs to move beyond comparative static assessments to consider dynamic processes of agriculture that reflect decisions from local to international scales.

(3) *Applied concepts of food system vulnerability.* The climate change community has often portrayed vulnerability as an outcome that results when adaptation is not sufficient to overcome negative consequences stemming from climate change (for example, see Ahmad et al. 2001).Vulnerability is primarily used to assess the severity of climate change issues. The natural hazards and famine research communities, however, have long viewed it as a property of socio-economic systems, reflecting inherent susceptibility and adaptive capacity based on determinants such as resource base, institutions, economy, and equity (for examples, see Davis 2002; Wisner et al. 2004). In this context, vulnerability is a reflection of the agricultural system itself, generated by social and economic resources, technology, and environmental constraints. This model of vulnerability as a dynamic, inherent property of agri-food systems provides a basis for understanding what is precipitating the vulnerability (e.g., limited managerial skills or domestic policy that disrupts economic or social safety nets) and offers a way to enhance the policy relevance of climate change/agricultural research. Identifying ways to enhance the adaptive capacity of agricultural systems is key to reducing vulnerability to climate change.

(4) *Enhanced science/policy linkages.* Climate change science has played a crucial role in advancing our collective understanding of how human activities have contributed to global change processes and has provided the foundation for many national and international policy efforts to reduce greenhouse gas emissions as a means to at least reduce the magnitude of global climate change. Efforts to mitigate the cause are without doubt part of the needed response to climate change and, to date, policy responses have focused largely on mitigation. Issues surrounding the adaptation of human systems (including agri-food systems) to climate change have received considerably less attention (internationally and in Canada) for many reasons, including relatively underdeveloped methods for assessing the capacity and likelihood of adaptation options to reduce food system vulnerabilities. The development of more robust and policy-relevant conceptual frameworks, and of analytical tools to estimate inherent food system vulnerabilities and the extent to which various adaptation options might reduce future vulnerabilities and enhance adaptive capacity, would provide a stronger foundation for developing a balanced climate change policy portfolio that addresses both mitigation and adaptation options.

These opportunities to broaden the climate change adaptation research in the agricultural sector are being realized in a number of ways. Several of the more recent scenario-driven studies have considered climate variability, focused on the farm scale, and explored relationships with non-farm forces. Other approaches (context-based and process-based) used in this book to present research on climate change adaptation and Canadian agriculture are introduced in subsequent chapters.

Acknowledgments
This review and appraisal has benefited from generous support from the Social Sciences and Humanities Research Council of Canada and from Procter & Gamble (Canada).

3
Context-Based Approach
Ellen Wall, Barry Smit, and Johanna Wandel

Another perspective for understanding issues related to climate change adaptation and agriculture focuses on the recognition that multiple factors influence adaptation possibilities and decisions. Factors such as market demands, regulation, and environmental conditions interact to influence sensitivity of the sector and the capability of individuals, communities, and regions to handle effects of climate change. These elements operate at several scales to shape the susceptibility of farming systems to climate stress and to determine the ability of those systems to adapt. The scope of context-based inquiry is usually quite broad, looking at larger socio-economic, political, and environmental features that constrain and/or increase the capacity for addressing climate change effects. Researchers working from this perspective often include material from the fields of environmental studies, political economy, risk assessment, and institutional analysis. This chapter reviews some insights regarding the broader context within which agricultural adaptation to climate change occurs, focusing primarily on climate risk perception and management and the role of various government programs in providing resources for meeting climate and weather challenges.

Risk Perception and Management
Managing risk is a key element of all businesses, and agricultural operations are no exception. Climate and weather risks are closely linked to management decisions regarding yield, input costs, and environmental factors (C-CIARN Agriculture 2003; Tyrchniewicz 2003). It is not surprising, therefore, that approximately half of Canadian producers surveyed in 2001 do not consider future climate change as having a separate, identifiable impact when asked: "What do you think the impact of climate change will be on Canadian agriculture?" (Aubin et al. 2003). Producers manage risks and pursue opportunities related to climatic and weather conditions as an implicit element of their business decision-making process.

Risk is commonly defined in terms of the probability of a consequence and its magnitude (Willows and Connell 2003); it encompasses both the likelihood that a certain state or event will occur and that it will have consequences of varying importance. Risk perception is usually considered a first step in risk management. Conventional wisdom suggests that before producers take steps to lessen the impacts of potential stresses, they need to be aware that such stresses exist. Research indicates, however, that the situation is not that straightforward, as the following examples illustrate.

Producers' perceptions of climate risk appear to vary by many characteristics of farm and farmer, including commodity produced (USDA 1999), and are related as much to economic, technical, and policy concerns as to climate. Reid and colleagues (2007) point out that southwestern Ontario producers are more likely to view climate as a risk if their operations are more vulnerable to it. Thus, in focus group sessions, cash crop producers, whose yields and incomes are more sensitive to climate, voiced more concern about the impacts of climate change than did livestock operators. Generally speaking, researchers note that Canadian producers think that the "industry" has provided and will continue to furnish adequate technological solutions to meet a variety of risks, including stress from climatic and weather conditions (Brklacich et al. 1997b; Bryant et al. 2000, 2004; Holloway and Ilbery 1996; Smit et al. 2000a).

Even though they may not explicitly acknowledge climate and weather risks as an important element of their management strategy, studies indicate that producers do take past performance (linked to agroclimatic conditions) into account when making production decisions. Working with producers in the South-West Montréal region of Québec, Bryant and colleagues (2004) found that although producers generally did not view climate change as an important issue, climate variability was intrinsically integrated into their decision-making process. Similarly, Smit and colleagues (1997) document close connections between corn hybrid selection and weather conditions the year prior to planting. Easterling (1996) reports that farmers in certain regions of the United States tend to recognize weather risks as substantial and modify their operations accordingly. Responses from central Canadian producers indicate a similar attitude. In this case, changing climatic and weather conditions (primarily increasing heat and dryness) led producers to adopt more conservation tillage, use different plant varieties and hybrids, and install irrigation systems, among other practices (C-CIARN Agriculture 2002).

Producers in the Prairie region also alter management practices to adapt to changing climatic and weather conditions. Some examples include increasing forage production, adding livestock to their farm operations, and introducing native grasses into grazing systems (C-CIARN Agriculture 2003).

According to Sauchyn (2003), more than 60 percent of Saskatchewan farmers surveyed indicated that they were preparing for climate change in response to current alterations in climatic and weather conditions. Sugar beet growers in the Prairie region have acted on their concerns about current risks from climatic and weather conditions by investing in research to improve production under persistent drought conditions. In particular, they have collaborated on research into the impact of short-term weather events on pesticide, herbicide, fungicide, and fertilizer applications, nitrogen management, harvesting and long-term storage of sugar beets after harvest, irrigation management, and increased conservation tillage.

Agricultural risk management strategies have been developed for many kinds of hazards and can be categorized according to four types: (1) risk reduction (e.g., diversification of the cropping system), (2) risk hedging (e.g., actions such as maintaining reserves), (3) risk transfer (e.g., using crop or other forms of insurance), and (4) risk mitigation (e.g., accepting a disaster payment from government sources) (Wandel and Smit 2000). The abilities of producers to manage risk depend largely on resources and the adaptation options available to them. Risk management takes place in light of the hazards associated with climate and in the context of socio-economic conditions and the influence from industry and government policies. Understanding the implications of various options is an increasingly important dimension of climate change research (MacIver and Dallmeier 2000).

The Role of the State and Industry in Climate Change Adaptation in Agriculture

Climate-related adaptation options for producers vary according to farm types and locations, and many options depend directly on government initiatives and programs, technology development, and financial opportunities beyond the farm gate (Dolan et al. 2001; Smit and Skinner 2002). For instance, a producer's ability to grow crops bred for climate-related traits, to implement more efficient irrigation systems, to manage soil and water resources more efficiently, and to diversify the business enterprises depends in part on available public and private sector services, technology, policies, and programs.

Despite the diversity in regional agricultural conditions and production, there are similarities in how producers view federal and provincial climate change adaptation policy (C-CIARN Agriculture 2004). It is evident that uncertainty and variability in all aspects of agricultural production present major risks that must be managed concurrently. Some producers want more government involvement, others want less. All appear to want stability, whether it is the stability of an insurance program or the stability of a not-rapidly-changing policy environment. At the same time, flexibility in policies and programs is crucial to ensure that diverse needs – from conditions in various types of commodity production, farming systems, biophysical environments, and personal circumstances – are met.

There are several ways that governments (often in partnership with the agri-food industry) can provide support for climate and weather risk management. These include sponsoring programs directed at building climate change adaptive capacity; supporting research programs that provide the necessary data and information for effective policy and program development; and ensuring that crop insurance and income stabilization are designed to be sustainable and effective.

Government Programs

There has been little analysis of the effectiveness of specific programs and their related incentives with respect to climate change adaptation in Canadian agriculture. In the American context, it has been argued that US farm policy works against producers' ability to adopt management practices (for example, switching crops and investing in water-conserving technologies) that constitute climate change adaptation (Lewandrowski and Brazee 1993). Leary (1999) investigated the cost/benefit issues related to climate change adaptation in the US and pointed out the high degree of uncertainty surrounding the benefits of adaptation. His recommendation is to delay implementing actions and programs that reflect only future climate concerns and focus instead on policy that generates benefits for current conditions (as long as they will also reduce vulnerability to future changes), which is now a widely promoted strategy in many areas. For example, Lewandrowski and Schimmelpfennig (1999) recommend that governments modify current farm programs regarding water conservation and restructure crop insurance and disaster relief to reflect added stress from altered weather and climatic conditions.

Representatives of the Canadian agri-food sector agree that climate change adaptation strategy is an implicit element in many existing environmental programs. For instance, in Atlantic Canada, there are now efforts to introduce techniques that will aid potato farmers in dealing with climate and weather risks. Better crop rotation and strip cropping, the use of winter cover crops and green manures, conservation tillage, residue management, and mulching have been encouraged. On steeper fields, the strategies include contour and cross-slope cropping, construction of diversion terraces and grassed waterways, enhancement of land drainage and nutrient management, and introduction of sediment control basins. In dealing with generic problems from runoff and erosion, many climate and weather risks are also addressed (Fairchild 2004).

Many government programs for the agri-food sector are administered in partnership with established agricultural organizations. Among these are commodity groups, federations, and associations devoted to specific concerns such as soil and water conservation. Collaboration with such groups on a variety of government programs encourages participation from their membership. In Canada, many of the greenhouse gas mitigation programs involving the agri-food sector and sponsored by federal and provincial governments have been integrated into existing initiatives, such as Greencover Canada (designed to improve grassland management practices, protect water quality, reduce greenhouse gas emissions, and enhance biodiversity and wildlife habitat), the National Farm Stewardship, and Environmental Technology Assessment for Agriculture (AAFC 2005). Existing programs related to crop production, variety development, land use, water resources, drought management, crop insurance, and safety nets represent means of implementing agricultural adaptation to climate change (Wall and Smit 2005).

Lee and colleagues (1999) point out the importance of evaluating options for farm-level adaptation within the larger environmental context. Their analysis of the US corn belt investigates the effects of winter cropping to reduce soil erosion. Despite the general benefit from this practice, the authors note that it can be a negative factor in some areas (i.e., those where the winter crop depletes soil moisture and subsequently adds to wind erosion and decreased corn productivity). Their findings support the view that government agencies need to consider all outcomes before implementing programs and policies that target specific production practices. Producers expressed a similar point of view regarding some Canadian agri-food sector policies (C-CIARN Agriculture 2003). For example, concerns were expressed regarding the conflict between promoting shelterbelts and regulations restricting pesticide use.

Canadian examples of programs related to production practices designed to reduce climate and weather-related risks at the farm level include promotion of conservation tillage systems, and various land and water resource management schemes. For instance, several initiatives, including the National Soil and Water Conservation Program (NSWCP) and the Agricultural Environmental Stewardship Initiative (AESI), have led to improvements in soil and water quality in many Canadian agricultural regions. Encouraging better land and water resource management (through conservation practices and groundwater protection) results in beneficial environmental conditions that also reduce negative impacts from climate-related events such as flooding and drought conditions.

Research Programs

Another way government and the agricultural industry work together to develop producers' capacity for managing climate-related risks is through joint research programs. Research efforts designed to improve agricultural sustainability are also relevant for climate and weather risk reduction and related adaptation options at the farm level. For instance, the Potato Research Centre in Atlantic Canada sponsors several projects aimed at reducing runoff and soil loss from high-intensity rainstorms (which are likely to increase with climate change). Rees and colleagues (2002) and Chow and colleagues (1999, 2000) document the effectiveness of hay mulching, different tillage systems, and grassed waterway systems, respectively, on land and water resources.

In the Prairies, crop diversification and irrigation research is supported by federal and provincial governments with the aim of improving agricultural sustainability in the region. An example is work completed at the Canada-Saskatchewan Irrigation Diversification Centre (CSIDC). This facility has sponsored at least twenty-six research projects and 106 field-scale demonstrations that have both direct and indirect implications for climate change (CSIDC 2000). Hogg and colleagues (1997), for example, examined the potential for using wastewater as an alternative source for irrigation systems. They concluded that the practice was acceptable as long as management practices were in place to offset potential problems with toxic compounds, infectious microorganisms, and salinity levels. Likewise, Zentner and colleagues (2002) at the Semiarid Prairie Agriculture Research Centre found that practices such as reducing summerfallow and adopting rotations of pulse crops and wheat for farms in semi-arid regions of Canada have both economic and environmental benefits. Findings such as these are relevant for adaptation to projected climatic conditions where an increase in dry spells is anticipated.

Similar examples of government and industry collaboration in research can be found in all Canadian provinces. Many research projects focus on improving technology so that agricultural production can be more efficient in light of increasing climate and weather risks. For instance, minimizing water use through highly efficient irrigation systems continues to be a major concern for researchers and producers in the Okanagan Valley of British Columbia. Projects out of the Pacific Agri-Food Research Centre include development of irrigation systems to deliver water precisely when orchard stock needs it (AAFC 2002).

Another example of a technological development that facilitates adaptation to altered climatic and weather conditions is crop breeding. There is little evidence, however, that crop developers emphasize "robustness" to climatic variations (also known as stability and resilience) in their programs (Rosenberg 1981; Smithers and Blay-Palmer 2001). It has been suggested that in the case of corn, such improvements might have been an "accidental" outcome related to the nature of breeding selection (Tollenaar et al. 1994;

Tollenaar and Wu 1999). On the other hand, van Herk (2001) notes that not only is climatic variability not a target for crop breeding (although it could be) but also an anomalous climatic season is seen as an inconvenience in field testing, with its results discarded, rather than an opportunity to test for and retain the robustness features of the crop variety.

Increasingly, crop development research is using biotechnological solutions; that is, genetically modified organisms are introduced into plant products. Without defining biotechnology, Evenson (1999) uses the IFPRI-IMPACT (International Food Policy Research Institute – International Model for Policy Analysis of Agricultural Commodities) model to support his claim that if genetic engineering is not incorporated into adaptation options, climate change will result in substantial crop losses and subsequent local food crises in the developing world. Agribusinesses (such as Performance Plants in Kingston, Ontario) are using genetic technology to develop high-yielding, drought-tolerant varieties of many commercially available crop species. This development is concurrent with research in other countries, such as the testing of genetically modified, drought-tolerant wheat in experimental plots by researchers in Mexico (CIMMYT 2004). Producer groups, however, such as Australia's Network of Concerned Farmers (2005), challenge claims that genetically modified crops perform better in drought situations.

The food safety aspects of biotechnologically developed products (for example, regarding potential problems with transgenic drought-tolerant rice) are a major concern for some scientists. In addition, experience with marketing challenges related to the use of biotechnological solutions makes some producers cautious about proceeding in that vein and points to the need for extensive research before proceeding too far. Easterling and colleagues (2004, 20) concur: "Social acceptability of technology given real or perceived risks can be a significant barrier to technological adoption and diffusion."

Crop Insurance and Income Stabilization
Uncertainty is a major feature for farming enterprises, and climate and weather variability compounds the instability in macroeconomic, biophysical, technological, and policy environments (Wandel and Smit 2000). One of the main strategies that producers employ to transfer risk from uncertain outcomes in the environmental and economic realms is to purchase crop insurance. The aim is to share the risk of yield loss, and therefore income loss, with a third party before or during exposure to risk. Federal and provincial governments play an important role in developing and supporting crop insurance schemes and have been involved with them for several decades (Asselstine 2003). Livestock insurance is also available but, for a variety of reasons, livestock plans are not as well developed as those for crops.

Figure 3.1

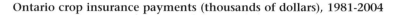

Ontario crop insurance payments (thousands of dollars), 1981-2004

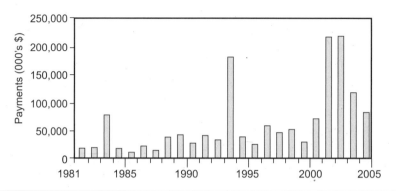

Source: Taken from total farm cash receipts, Statistics Canada, 1981-2004.

National crop insurance statistics indicate increased use of existing pro-grams in recent years. Statistics Canada (2003) reports that the dramatic increase in producers' receipts from program payments in the first six months of 2003 (79.2 percent increase from the same period of 2002 and almost double the previous five-year average) is largely a result of record payments through crop insurance programs following two consecutive years of drought for Prairie producers.

Saskatchewan alone paid $825 million dollars for 2002 crop losses, a fig-ure more than double the amount needed for the 1988 drought. Alberta has spent over $1.8 billion on ad hoc drought relief since 1984 (SSCAF 2003). Stress on government-supported insurance programs is not confined to Prai-rie provinces, as data from Ontario crop insurance payments reveal a simi-lar pattern (see Figure 3.1).

Given their increasing importance to the agri-food sector, crop insurance issues are well researched in the agricultural economics and policy litera-ture. Insurance claims are linked to yield loss, and as yield is closely tied to climatic and weather conditions, most crop insurance studies have some degree of relevance for climate and weather risk transfer. Some, however, examine crop insurance issues specifically in light of changing climatic and weather conditions. It has been argued that insuring against the cause of crop damage (e.g., adverse climate and weather events) is more effective than insuring against crop loss (Turvey 2001). Crop insurance addresses low-risk, high-probability events, whereas most non-agricultural insurance products are designed to manage high-risk, low-probability events such as floods or fire (Chance 2003). If payouts are based on objective weather data, the need for insurance adjusters to assess crop damage is reduced, thereby lessening costs and subjectivity. In Alberta, however, there is evidence that the system might be problematic. Farmers in some regions claim that moisture data used by the Alberta Financial Services Corporation include inaccurate figures suggesting that moisture was above normal levels in the summer of 2003, even though their fields are drier and more grasshopper-infested than ever before.

Mahul and Vermersch (2000) analyze the problem of hedging crop risk against crop yield insurance futures and options rather than weather derivatives. They conclude that catastrophic weather events have become the major factor in yield variability and therefore crop insurance payments. Ker and McGowan (2000) investigate implications of "weather-based adverse selection" in the United States. Adverse selection occurs when growers of different loss-risk are charged premiums that do not reflect this difference. As a result, only those taking on high risks buy insurance and the insurer becomes more likely to incur actuarial losses. Adverse selection generally exists whenever the insured person has better knowledge of the relative "riskiness" of a particular situation than the insurance provider does.

Ker and McGowan (2000) also note that farmers in the US tend not to include weather-based information in their decisions on whether to purchase insurance, but private insurance companies do (in pursuit of reinsurance) and subsequently benefit from government compensation under current reinsurance agreements. Farmers' use of crop insurance in Canada has some similarities to that in the United States. Smithers and Smit (1997b) suggest that crop insurance programs appear to reduce the sensitivity of producers to unfavourable years in that producers make decisions knowing that the insurance program assures some income. In general, there may be less incentive to adopt individual risk management strategies if governments serve that purpose (Lewandrowski and Brazee 1993; Smit et al. 2000a).

Additional strategies for managing climate and weather risks include risk mitigation initiatives such as income stabilization programs and disaster relief. These programs provide payouts during particularly bad years. The future of such programs in Canada is negotiated between federal and provincial governments. The Business Risk Management section of the Agricultural Policy Framework (APF) now provides compensation for unfavourable consequences from climatic and weather conditions through existing programs for income stabilization and disaster relief. Policy makers and producers alike recognize the need to include other climate-related adaptation strategies in additional "pillars" of the APF (food safety, innovation, environment, and renewal) (C-CIARN Agriculture 2004).

This chapter has focused on the context-based research approach for climate change adaptation and agriculture. The research examines climate risks and adaptations in the context of existing risk management strategies, including adjustments at the farm level, research into new cultivars, and participation in institutional programs. Studies following a context-based perspective reveal that management of climate risks is influenced by other forces, and is often undertaken in light of these. Climate and weather risk management cannot be understood as an isolated component in farm decision making, policy development, and research. Adaptation is rarely a response to climate change alone, but reflects a process that is driven by minimization of economic risk, incorporation of technological advances, and use of a number of industry- and government-supported programs. Specific examples of context-based research are presented in Part 3.

4

Process-Based Approach

Johanna Wandel, Ellen Wall, and Barry Smit

The process-based approach focuses on producers' adaptive decision making and documents knowledge of actual adaptive behaviour, mainly at the farm level. This perspective also focuses on factors that drive or constrain adaptation, providing the basis for assessing future adaptive capacity to handle risks (and opportunities) from climate change. Research that focuses on the practical applications and processes of adaptation is still not common, at least not under the label of "adaptation" research, and not in the climate change field. However, a substantial body of scholarship relevant to the process-based approach does exist in the fields of resource management, community development, risk management, planning, food security, livelihood security, and sustainable development (Alwang et al. 2002; Gittel and Vidal 1998; Haimes 2004; Sanderson 2000). Research examining agricultural adaptive capacity for dealing with climate change draws from many of those sources and is directly related to "vulnerability assessments," which are prominent among process-based research because of their emphasis on bottom-up, stakeholder-driven, applied research.

Process-based approaches focus on the need for, and mechanisms of, adaptation in a particular region or community in order to identify the means of implementing adaptation initiatives or to enhance adaptive capacity. The purpose is practical – to characterize vulnerability and adaptive capacity in order to initiate adaptive measures or practices, or in order to improve the adaptive capacity of that particular community. Emphasis is on the ways in which the system or community experiences changing conditions and on the processes of decision making in this system that may accommodate adaptations or may provide means of improving adaptive capacity (Ford and Smit 2004; Keskitalo 2004; Lind 2003; Sutherland et al. 2005).

Given the relatively recent interest in using a process-based approach for understanding climate change adaptation and agriculture, the number of studies for review in this chapter is limited. The discussion does provide information on basic terms, conceptual frameworks, and methodology pertinent for process-based analysis.

Concepts, Definitions, and Frameworks

Adaptation in the context of human dimensions of global change usually refers to a process, action, or outcome in a system (household, community, group, sector, region, country) that enables that system to cope with, manage, or adjust to some changing condition, stress, hazard, risk, or opportunity (Smithers and Smit 1997a). It is possible to distinguish adaptations according to the degree of adjustment or change required from (or to) the original system (Risbey et al. 1999). For instance, a simple adaptation might be to use more drought-resistant cultivars for an agricultural system facing water shortages. A more substantial adaptation might be to shift away from crop farming to livestock production based on intensive grazing with native grasses (C-CIARN Agriculture 2003).

The ability to undertake adaptations is referred to as *adaptive capacity*. Adaptation and adaptive capacity are relative concepts – adaptation *of* something *to* something; the capacity to act *in light of a* stress or suite of stresses. *Exposure-sensitivity* refers to the susceptibility of a system to stresses and the degree to which the system is affected by stresses. These concepts of adaptive capacity and exposure-sensitivity together comprise the *vulnerability* of the system. "Vulnerability" refers to the susceptibility of a system to stresses, reflecting both the degree to which the system is sensitive to hazardous effects (exposure-sensitivity) and the ability to cope or adapt (adaptive capacity) (Smit and Wandel 2006).

Applications of these terms range in scale from the vulnerability and adaptation of an individual or household to a particular climatic stress such as drought, through the vulnerability and adaptation of a community to various environmental stresses, to the vulnerability of humankind (or the global ecosystem) to all stresses and forces.

A conceptual model of vulnerability (Figure 4.1) has emerged in the climate change scholarship (Downing 2001; Kelly and Adger 2000; Smit and Pilifosova 2003; Yohe et al. 2003). Consistent throughout the literature is the notion that the vulnerability of any system (at any scale) is reflective of (or is a function of) both the exposure and/or sensitivity of that system to hazardous conditions and the ability or capacity or resilience of the system to cope, adapt, or recover from the effects of those conditions. These concepts are labelled in different ways and given different emphases in various fields.

In agriculture, a crop-based farming system that is exposed to drought (the hazard) will be susceptible to yield declines and economic loss (sensitivity), which can be moderated by adaptive strategies such as irrigation,

Figure 4.1

Conceptual model of vulnerability

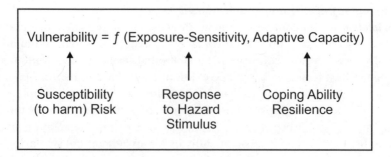

Source: After Smit and Pilifosova (2003).

crop insurance, the use of less moisture-reliant cultivars, and the farmer's management ability. The availability and use of such strategies reflect the system's adaptive capacity. Thus, a modest change in climate may have little direct effect on a system that is not highly sensitive to the hazard (e.g., feedlot operations that do not grow their own feed) or a system that is highly adaptable (e.g., farmers who carry sufficient crop insurance or who have irrigation capacity). The same change, however, may have a substantial effect on a system that has both high sensitivity and low adaptability (e.g., a cash crop operator without access to irrigation, facing low commodity prices, and not carrying crop insurance).

A related concept, and one of prime importance for agricultural production under a changing climate, is a system's "coping range." This term is defined by the range of conditions within which a system can function, accommodate, adapt to, and recover from (de Loë and Kreutzwiser 2000; Jones 2001; Smit et al. 2000b; Smit and Pilifosova 2003). Canadian agriculture has historically coped with (or adapted to) normal climatic conditions and deviations from the mean to some degree. (For instance, see Chapter 6 for details on agricultural adaptation in the Canadian Prairies.) Exposures involving extreme events that lie outside the coping range may exceed the adaptive capacity of the farm or community, however (see Figure 4.2).

A system's adaptive capacity or coping range is not static. Coping ranges are flexible and respond to changes in economic, social, political, and institutional conditions over time. For instance, soil erosion may gradually reduce a farming system's coping ability and narrow its coping range. At the same time, the farmer may have increased financial resources from non-farm income, adopt technological changes that enable increased yields, or diversify operations into non-crop production, all of which may lead to an

Figure 4.2

Coping range and extreme events

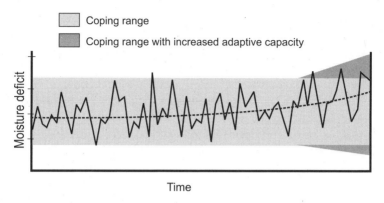

Source: After Smit and Pilifosova (2003).

---

increase in adaptive capacity (de Vries 1985; Folke et al. 2002; Smit and Pilifosova 2003).

Figure 4.2 provides an illustration of how a system's coping range reflects adaptive capacity relative to exposure, and how these relate to vulnerability. In this case, the agricultural community is sensitive to changes in moisture. Moisture deficits and the occurrence and severity of drought conditions vary from year to year, yet the system depicted in Figure 4.2 is able to cope with a degree of variation around the mean or average conditions. The amount of variation the system can deal with is indicated in the shaded area, referred to as the "coping range" (which could also be called the "adaptive capacity thresholds" of the system). As mean moisture deficit increases, however (as is expected with climate change for some Canadian agricultural regions), the entire inter-annual moisture distribution shifts and the system will experience, and be more exposed to, an increase in the frequency and magnitude of events beyond the coping range. If everything else were to stay the same, this would increase the vulnerability of this system. To the extent that the system may be able to expand the coping range or enhance its adaptive capacity to deal with these exposures, it will reduce its vulnerability to drought risk. The forces that influence the ability of the system to increase adaptive capacity are the driving forces, external factors, influencing processes, and determinants of adaptive capacity (Adger et al. 2004; Blaikie et al. 1994; Kasperson and Kasperson 2001; Turton 1999; Walker et al. 2002; Wilbanks and Kates 1999).

In agricultural operations, the coping range or ability to undertake adaptations is widely understood to be dependent on or influenced by a variety of conditions, including managerial ability; access to financial, technological, and information resources; infrastructure; the institutional environment within which adaptations occur; political influence; and social networks, among other factors (Adger 1999; Adger et al. 2001; Blaikie and Brookfield 1987; Hamdy et al. 1998; Handmer et al. 1999; Kelly and Adger 2000; Toth 1999; Watts and Bohle 1993; Wisner et al. 2004). The determinants of adaptive capacity are not independent of each other. For example, for farm families under stress, the presence of strong social networks could allow greater access to economic resources, expertise, supplementary labour, and psychological support.

The process-based approach identifies the features and conditions that make up the broad elements in Figure 4.1, and characterizes the processes that contribute to vulnerability. Several distinctive features typify research conducted with a process-based approach:

- Experience and knowledge of community members are used to characterize pertinent conditions, community sensitivities, adaptive strategies, and decision-making processes related to adaptive capacity or resilience.
- The motivation is to identify what can be done, in what way, and by whom, in order to moderate the vulnerability to the conditions that are problematic for the community (Pahl-Wostl 2002; Moss et al. 2001; Morduch and Sharma 2002).
- Risks (and opportunities) associated with climate change (or other environmental changes) are addressed in decision making at some practical level. Adaptive responses are rarely, if ever, made in light of climate change alone (Handmer et al. 1999; Huq and Reid 2004; Huq et al. 2003; Morduch and Sharma 2002).

Research following the process-based approach results in distinctive and practical information about adaptive capacity generated from applying these three characteristics.

Assessing Adaptive Capacity for Climate Change Impacts

A distinguishing feature of process-based approaches for understanding climate change impacts and adaptation is that researchers do not presume to know the exposure-sensitivities and determinants of adaptive capacity in the community of interest. Rather, these are identified after working with members of the community itself. This endeavour requires the active engagement of stakeholders and considerable effort to ensure legitimacy within the community. There can be major challenges for information collection and the integration of information from multiple sources.

Chapter 1 points out that a process-based methodology is associated with social sciences and participatory, "bottom-up," experience-based assessments of community conditions. Several disciplines and fields of inquiry (anthropology, sociology, ethnography, geography, risk assessment, rural development, international development, and food security) employ such methods (Bollig and Schulte 1999; Pelletier et al. 1999; Ryan and Destefano 2000; Smith et al. 2000). In the fields of climate change adaptation and disaster management, analytical frameworks featuring versions of vulnerability assessments have been developed and some have been applied (Jones 2001; Lim and Spanger-Siegfried 2004; Schröter et al. 2005; Turner et al. 2003).

Vulnerability Assessments

The starting point for vulnerability assessments is the system of interest (for instance, agricultural production in a specific region) without a priori hypotheses of relevant exposure-sensitivities or determinants of adaptive capacity. Researchers must be flexible enough to recognize multiple sources of capacity, including political, cultural, economic, institutional, and technological factors. Furthermore, a vulnerability assessment needs to allow for the interaction of various exposure-sensitivities and adaptations and adaptive capacities over time, since what is vulnerable in one period is not necessarily vulnerable (or vulnerable in the same way) in the next, and some exposure-sensitivities (e.g., those recognized as "creeping hazards" by Wisner et al. 2004) develop slowly over time. Vulnerability assessments must also recognize sources of exposure-sensitivities and adaptive capacities at various scales, from the individual to the national.

A conceptual model of a vulnerability assessment includes basic elements and defining features (see Figure 4.3).

Figure 4.3

Conceptual framework for vulnerability assessment

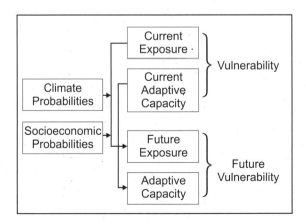

Source: After Ford and Smit (2004).

The identification of *current exposures* involves documenting the conditions or risks that people have had to deal with or are dealing with in their lives, livelihoods, businesses, sectors, and so on. Researchers complete this stage using ethnographic in-community fieldwork, relying on research tools such as semi-structured interviewing and participant observation, as well as insights from local and regional decision makers and resource managers. Both physical and non-physical stresses as well as the community characteristics (livelihoods, settlement, etc.) contributing to problematic conditions are identified. In essence, researchers are guided by the following questions:

- What sorts of conditions have posed problems for this person, household, group, community, or institution in the past?
- What problems are currently being dealt with?

Assessing *current adaptive capacity* involves identifying the ways in which the community deals and has dealt with exposures. The goal here is to answer the following questions:

- How have individuals, households, groups, and institutions dealt with, coped with, managed, or adapted to the problematic conditions of the past?
- What adaptations or adaptive strategies were employed? How, why, by whom, and under what circumstances?
- How are problematic conditions currently being dealt with?
- How effective or otherwise are current strategies?
- What are current barriers and enabling factors for management of problematic conditions?

Assessing current adaptive capacity requires an identification of the broader conditions that constrain or facilitate adaptive initiatives. This introduces the following questions:

- If there have been measures or policies or other forms of institutional support that helped deal with exposures, what were the conditions that made them feasible and effective?
- Are there needs and opportunities not realized?

Not all adaptive strategies will be at the scale of the community since some are institutionalized at a broader level and beyond the perceived or real control of community members. This reality influences how responses to the questions are followed up.

Together, the above questions represent the characterization of current exposures and adaptive capacities and provide some sense of the system's ability to address changing conditions and risks in the future.

Once current exposure and capacity have been documented, it is possible to move on to the next stage, that of assessing future vulnerability (Figure 4.3). *Future exposures* relate to conditions that are expected to be risks or opportunities to the community at a later date. In this case, researchers, using a variety of methods, attempt to assess the likelihood that there will be problematic or positive conditions in the future. Modelling of the climate/weather and an array of related environmental conditions (such as moisture levels) that have been identified as relevant will provide a valuable resource for assessing future exposure. In addition, trend analysis in economics, policy, demographics, and other social variables will give some indication of future exposure to both positive and negative conditions in the socio-economic realm.

An agricultural example of future exposure may take the following form. An assessment of current exposures reveals that a specific agricultural community has, in the past, been unable to provide adequate water resources for production. At present, agriculture is a dominant economic activity in the community, and this is encouraged by state-level food security policies. Climate and hydrologic modelling may reveal a likelihood of diminishing water supplies, while demographic analysis reveals a classic "pyramid" age/sex distribution indicating rapid future population growth. Policy analysis shows little indication of a change in the institutional encouragement of agriculture and a strong lobby to maintain existing water market control structures relying on prior rights. Thus, researchers can conclude that future exposure will be greater than current exposure due to the combination of these factors.

Analyzing *future adaptive capacity* follows and focuses on understanding the manner and degree to which current management practices could deal with or accommodate the estimated future exposures. More broadly, the assessment could consider the degree to which community members have adaptive capacity in terms of the scope, resilience, resources, and potential to deal with expected future exposures. Thus, in our agricultural example above, the recognition that there is going to be a substantial problem related to water resources in the future would be a signal for government, industry, and residents to take action. A number of elements would form the basis for future decisions about policy, programs, and/or regulation. All of them would have to reflect a major reduction in water availability, and might include such initiatives as restrictions on human settlement, investment in high-efficiency water systems, enforced regulations regarding water use, and substantial training and education for improved water resource management.

This chapter has provided a basic description of the process-based approach for research on climate change adaptation and agriculture. Specific examples of employing such a perspective are found in Part 4.

Part 2:
Impact-Based Studies

The three chapters in Part 2, each with a specific regional focus, present studies on climate change adaptation in Canadian agriculture that employ to some degree the impact-based approach. Samuel Gameda, Andrew Bootsma, and Daniel McKenney provide a synopsis of impact-based research on crop production in eastern and central Canada in Chapter 5, while David Sauchyn focuses specifically on the Prairie regions in Chapter 6. Impact-based assessments build on climate change scenarios for British Columbia form the basis for Chapter 7, where Denise Neilson and colleagues provide details from their work in the Okanagan region.

5
Potential Impacts of Climate Change on Agriculture in Eastern Canada
Samuel Gameda, Andrew Bootsma, and Daniel McKenney

In this chapter, the focus is on particular impacts from changes in agro-climatic indices that may affect crop production in eastern and central Canada. Various agroclimatic indices have been used in the past to assess the climatic conditions for production of cultivated crops in Canada. Growing degree days above 5°C (GDD) are commonly used as an index for general plant growth and development, although the base temperature will vary for specific crops (Chapman and Brown 1978; Edey 1977). The period for accumulating GDD is normally considered to be the growth period for perennial forage crops. Water deficits (DEFICIT, defined as PE – P, where PE and P are the seasonal potential-evapotranspiration and precipitation, respectively) and effective growing degree days above 5°C (EGDD) have been used as principal climatic variables to rate the climatic suitability of land for production of spring-seeded small grains (Agronomics Interpretations Working Group 1995). EGDD are similar to GDD, except that they are adjusted for the effect of longer daylengths on spring cereals at latitudes north of 49°N and are accumulated over a shorter time period than GDD. The aridity index (ARIDITY) has been used as a primary classifier in the Soil Water Regime Classification System for Canada (Expert Committee on Soil Survey 1991). This index is similar to irrigation requirements determined by a daily water budgeting technique in which irrigation water is added when the soil moisture drops to 50 percent of the available water capacity (AWC) (Shields and Sly 1984). Crop heat units (CHU), also known as corn heat units, have been widely used to rate the suitability of various regions for the production of corn and soybeans (Bootsma et al. 1992, 1999; Brown and Bootsma 1993; Chapman and Brown 1978; Major et al. 1976). These indices were adopted for our research because of their important influence on crop performance and common acceptance in the past as indicators of crop suitability.

Results from three studies form the basis for this chapter. The first study generates agroclimatic indices (heat units and water deficits) for the Atlantic region of Canada for a baseline climate (1961-90 period) and for two future time periods (2010-39 and 2040-69). Climate scenarios for the future periods are based primarily on outputs from the Canadian General Circulation Model that included the effects of aerosols (CGCM1-A), but variability introduced by multiple Global Climate Model experiments is also examined. The authors then discuss the implications of their projections for annual crop production in the Atlantic region.

In the second study, agroclimatic indices (heat units, aridity index) in agricultural regions of Ontario and Québec are determined for the baseline (observed) climate (1961-90 average) and for three future time periods (2010-39, 2040-69, and 2070-99). Climate scenarios for the future time periods are based on the output of eleven GCM simulations, using GCMs developed in Canada, the United Kingdom, and Germany. All GCM simulations used in this study assumed greenhouse gas (GHG) forcing based on the IS92a emission scenario developed by the Intergovernmental Panel on Climate Change (IPCC 2001).

The third study focuses on the impacts of climate change on winter survival of perennial forage crops in eastern Canada. In this study, causes of damage during fall, winter, and spring were identified and agroclimatic indices expressing the relative intensity of each cause were developed. Fall indices reflect the effects of temperature and precipitation during cold hardening. Winter indices integrate the interactions between cold intensity, cold duration, and the protective role of snow cover; assess the loss of cold-hardiness due to warm temperatures; and estimate the potential damage to roots by soil heaving and ice encasement.

These three studies provide classic examples of the impact-based approach as described in Chapters 1 and 2. The balance of this chapter is devoted to reporting in some detail the procedures, results, and implications from the first two studies. For the third study, only the major findings are briefly summarized.

Study 1: Agroclimatic Indices and Annual Crops in the Atlantic Region of Canada

This investigation[1] included the Atlantic provinces (Figure 5.1), where the climate is predominantly a modified continental type, moderated somewhat by proximity to the Atlantic Ocean (Chapman and Brown 1978). Winters tend to be cold and snowy, while growing season conditions are generally moderately cool with ample moisture. The main agricultural areas of this region are also identified in Figure 5.1.

Baseline Climate and Climate Change Scenarios

The baseline climate data used for this study are thirty-year monthly climate normals for the 1961-90 period (Environment Canada 1994) for daily

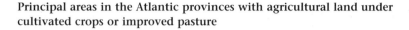

Figure 5.1

Principal areas in the Atlantic provinces with agricultural land under cultivated crops or improved pasture

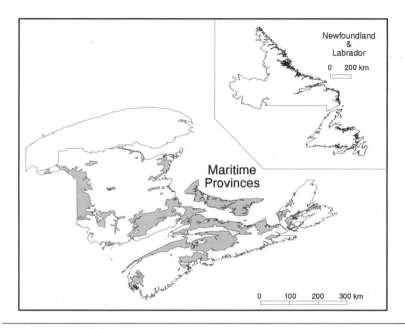

maximum and minimum air temperatures and total precipitation. Monthly mean values of each variable were interpolated to a grid of 500 arc seconds (approximately 10-15 kilometres), using Digital Elevation Model (DEM) data and a thin plate smoothing spline surface fitting technique (Hutchinson 1995). Interpolations were made using a software package called ANUSPLIN (Australian National University SPLINe software) (Hutchinson 2000). Surfaces were created using trivariate splines with latitude, longitude, and elevation as the independent variables. (Complete details for the interpolation procedures used are available in McKenney et al. 2001.)

Climate change scenarios for temperature and precipitation for the periods 2010-39 and 2040-69 were constructed, based on the output of the first-generation coupled Canadian General Circulation Model (CGCM1-A) (Boer et al. 2000). In the CGCM1-A scenario, the GHG forcing increase is equivalent to a doubling of carbon dioxide by around 2050 compared with the 1980s. Using only one GCM scenario in our study for estimating the potential effects of climate change is not ideal, but it does provide an indication of one plausible outcome.

Mean monthly differences in temperature between the GCM simulations (for each of the two future time periods) and the historical simulation of the GCM for the 1961-90 period were interpolated from the coarse GCM grid to the finer regional grid scale using the thin plate spline methods of ANUSPLIN. For precipitation, the ratios of the simulated GCM values for the future time period over the simulated baseline period were interpolated.

Temperature differences and precipitation ratios were applied to the gridded mean monthly maximum and minimum air temperature and precipitation data, respectively, for the 1961-90 period to construct gridded data for the two future time periods. While the application of more sophisticated downscaling techniques, such as statistical downscaling (Wilby and Wigley 1997), is advantageous for producing scenarios, particularly within the land/sea interface of this region, the simpler interpolation procedure was used here to provide one plausible outcome of climate change in the region. GCM outputs are not considered to be precise forecasts of future climate but rather only plausible scenarios of what could happen in the future given a trajectory of economic activity and population levels (IPCC 2001). The gridded scenario data generated in our study are available from the CCIS (Canadian Climate Impacts Scenario Project) website (Canadian Institute for Climate Studies 2002).[2]

Interpretation and Implications

Average CHU for the baseline climate are typically in the 2,400-2,600 CHU range in the main agricultural areas in our study. These increase to 2,600-3,000 CHU for the 2010-39 period, and to 3,000-3,200 CHU for 2040-69. The increases are in the 300-500 and 500-700 CHU category, respectively, for the two future time periods (Figure 5.2).

Patterns of changes in EGDD are very similar to the pattern of change for CHU. Consequently, it is estimated that by 2040-69, EGDD in the main agricultural areas will range from 1,800 to over 2,000 units. Only the eastern part of Labrador indicated a decrease in EGDD.

Water deficits (DEFICIT) for the baseline period varied from over 100 millimetres in central New Brunswick to a surplus (negative value) in excess of 100 millimetres in the more humid regions, such as southwestern tip of Nova Scotia, Cape Breton Island, and southeastern parts of Newfoundland and Labrador. Patterns for deficits/surpluses were similar for the future time periods. The greatest increases in deficits were in New Brunswick, with a large region in the interior indicating increases in the 25-50-millimetre class for the 2040-69 period (Figure 5.3). In areas of eastern Nova Scotia and in much of Newfoundland, precipitation surpluses increased by 25-50 millimetres for this time period.

The climate changes expected to occur in the Atlantic region within the next fifty years or so, based on GCM experiments and a "business as usual" emission scenario for GHGs, are likely to have significant impacts on crop production in the region. Based on our scenario of increased availability of

Figure 5.2

Changes in crop heat units (CHU) for two future time periods compared with the 1961-90 baseline period for the Atlantic region

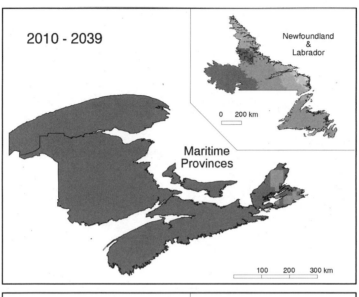

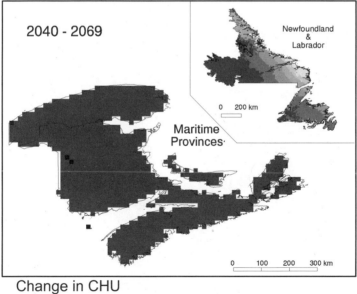

Figure 5.3

Change in water deficits (mm) for two future time periods compared with the 1961-90 baseline period for the Atlantic region

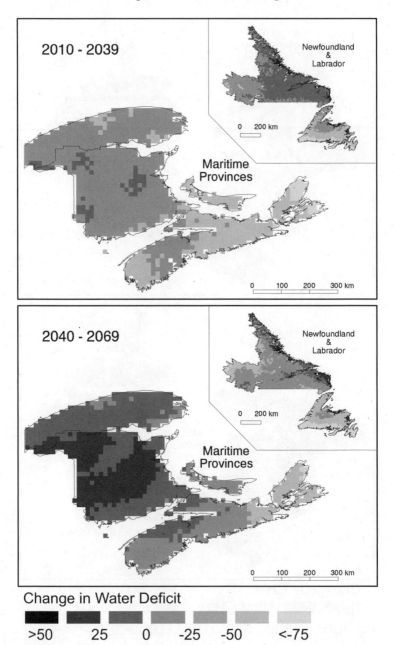

CHU with only slight to moderate changes in water deficits, we would expect increases of about 54 percent and 29 percent in average yields of corn and soybeans, respectively, for the 2040-69 period compared with the mid-1990s, in areas where corn is presently grown (Bootsma et al. 2005b). This does not take into consideration the potential effects of technological and genetic improvements, or potential changes in the impacts of weeds, pests, and diseases on yield. Production of corn and soybeans would likely expand into areas of the Atlantic provinces that are presently unsuitable because of limitations imposed by the climate.

Yields of barley are likely to change only slightly, as shifts towards earlier planting and ripening dates would likely result in little change in overall exposure to heat units (EGDD). Based on our climate scenario, we estimate that increased EGDD will likely reduce average yields of barley slightly, but that this decrease will be more than offset by the direct effect of higher CO_2 concentrations, so that, overall, average yields will increase by about 15 percent by the 2040-69 period. Since the competitive advantage of barley in relation to corn and soybeans will be reduced, this will likely lead to significant reduction in land area seeded to barley and increases in corn and soybean acreage. Overall, the crop productivity will be increased by increased yields and the switch to high-energy and high-protein-content crops (corn and soybeans) that are better adapted to the warmer climate. There will likely, however, still be a considerable area of land seeded to barley and other small grain cereals, as these are very desirable in rotation with potatoes. The potential impact of these possible shifts in crop production on soil erosion in the region needs to be evaluated. Land-use changes may become a significant issue as forestry is currently the dominant land use in much of the region.

The certainties in our yield and production scenarios are limited, since not all possible factors were included in our study. The estimates are based primarily on the effect of increasing temperatures on average yield as determined from field trials conducted under existing climates. Numerous factors – such as pests, diseases, direct effects of CO_2, plant breeding, soil fertility, management practices, and socio-economic factors – could affect future crop production under a changed climate. Additional research is needed to determine the potential effects of these factors in a more rigorous fashion. Effects of climate change on other crops, such as potatoes, forages, and winter wheat will also have an impact on cropping decisions by producers. In the case of winter wheat, overwintering conditions are also extremely important, as increased potential for winter injury under climate change could limit production of winter wheat in the region (Bélanger et al. 2002).

Adaptive responses to climate change by producers would also affect future changes in crop production. For example, if producers delay their selection of hybrids/varieties that are best adapted to changed climatic conditions, potential gains in crop production would be delayed. The potential for increased crop production would be realized earlier, however, if confidence in long-term climate forecasts improves and producers adopt a proactive stance in adapting to climate change.

Study 2: Agroclimatic Indices and Annual Crops in Regions of Ontario and Québec

The second study[3] of climate change and annual crop production focused on regions in Ontario and Québec where land is suitable for agriculture. The area may be classed as a humid continental-type climate, although there is considerable variation over the agricultural regions, with warm summers and moderate winter temperatures in southern areas and moderate summer temperatures and cold winters in more northern regions. In southwestern Ontario, temperatures are significantly moderated by the Great Lakes and the climate is more humid than in northern regions. On average, precipitation is remarkably uniform throughout the year, although there is a tendency towards somewhat lower precipitation during winter months, particularly in more northern agricultural regions (Chapman and Brown 1978).

Baseline Climate and Climate Change Scenarios

The baseline climate data used for this study was a thirty-year monthly mean of average daily maximum and minimum air temperatures and total precipitation for the 1961-90 period interpolated from available station data (Environment Canada 1994) to a grid of 500 arc seconds (approximately 10-15 kilometres), as described in Study 1.

Average monthly differences in mean maximum and minimum air temperatures and percent of precipitation between three future time periods (2010-39, 2040-69, and 2070-99) and the 1961-90 values were extracted from the CCIS website for eleven different GCM model simulations. Model selection was based on providing a suitable range in the data, and having available all three variables for the three future time periods. The monthly average changed fields (difference values for temperature, percent of 1961-90 precipitation converted to ratios) from each GCM experiment were interpolated to the 500 arc second grid using a procedure similar to that described for Study 1.

Interpretation and Implications

Average CHU (Figure 5.4) are projected to increase substantially with time over all of the study area due to longer and warmer growing seasons. These increases suggest that areas that are presently marginal or unsuitable for grain corn and soybean production due to lack of heat (i.e., areas with less than 2,300 CHU) will become suitable for production of these crops sometime in the future. Also, producers in areas presently well suited to grain

Figure 5.4

Crop heat units (CHU) in main agricultural regions of Ontario and Québec

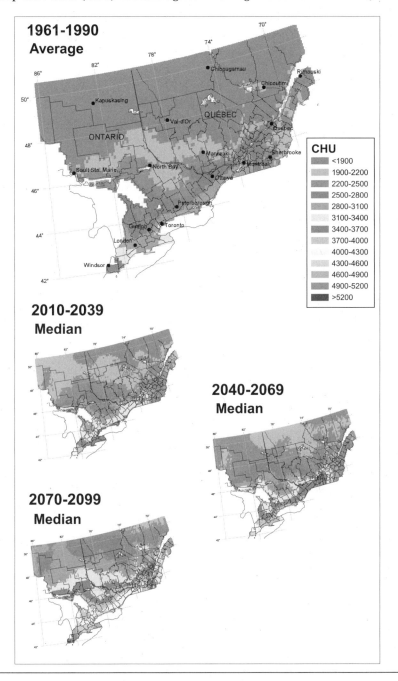

corn and soybean production may be able to select longer-season hybrids/ varieties, which will potentially increase yields by taking advantage of the more favourable thermal regime.

Patterns in increases in GDD for future time periods are very similar to those for CHU. A large portion of the land area of Ontario and Québec that has soils that are suitable for cultivated field crops (Canada Land Inventory 1975) is rated in the range of 1,750-2,250 GDD for the baseline period. Median values for these same regions are typically in the range of 2,250- 3,000 GDD for the 2040-69 period and 2,750-3,500 GDD for the 2070-99 period. The growing season length (GSL) for perennial forage crops is de- fined by the period over which GDD above 5°C are accumulated. This length is anticipated to increase by about 30-45 days for the 2070-99 period (Figure 5.5).

The patterns of change anticipated for EGDD are very similar to the GDD patterns, although the absolute values are somewhat lower because of the shorter growing season length for EGDD. As a result of changes in starting and ending dates for accumulating EGDD, the length of the growing season over which EGDD are summed (GSL 2) is typically increased by about 40-55 days by the 2070-99 period.

The potential changes in the ARIDITY index are shown in Figure 5.6. The values shown are for the extreme 10 percent probability level (i.e., values that are exceeded only about one year in ten). The AWC was assumed to be constant (150 millimetres). If actual AWCs were used, the spatial patterns would be considerably more complex than shown. Results indicate an in- crease of irrigation requirement that is exceeded one year in ten of about one ARIDITY class (50 millimetres) in most areas by 2070-99.

Projected changes in climate based on GCM scenarios have significant im- pact on agroclimatic indices that are important to agriculture in Ontario and Québec. This is particularly true of heat-related indices such as CHU and GDD, as these are expected to increase under warmer climates. Increases in CHU will result in higher potential yields of corn and soybeans, provided that adequate moisture is available for plant growth. Data from the US corn belt suggest, however, that corn yields would reach a plateau or even decline above 3,500 CHU as moisture may become more limiting (Bootsma 2002). Increased GDD combined with an extended growing season length could boost potential yields of forages such as alfalfa, by allowing for additional cuts (Bootsma et al. 1994), although the ability to achieve these higher yields may be limited by changes in moisture supply and winter survival. Yields of spring-seeded small grain cereals are less likely to be affected positively by the anticipated increases in EGDD, as shifts towards earlier planting and ripen- ing dates would likely result in little change in overall exposure to heat units.

Indices related to moisture balance suggest that some increases in soil aridity are likely for future periods, and this may place some constraints on

Figure 5.5

Growing season length (GSL) for accumulating growing degree days (GDD) in Ontario and Québec

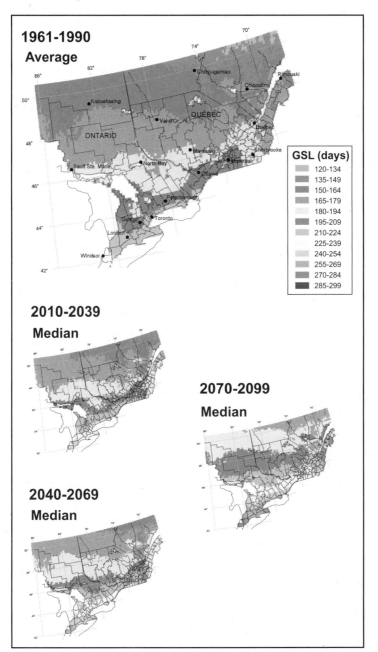

Figure 5.6

Aridity index for soil with 150 mm AWC (10 percent probability level) in Ontario and Québec

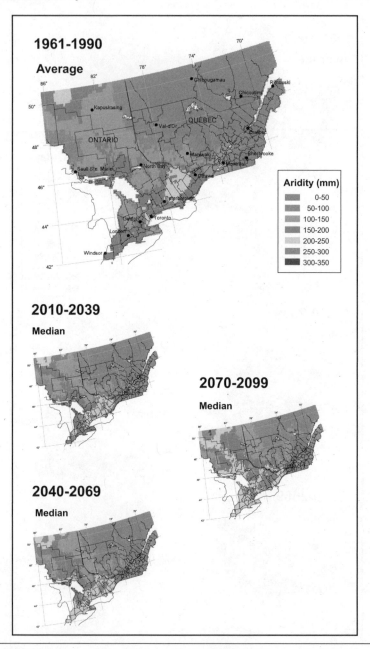

the increased yield potential expected for some crops as a result of longer, warmer growing seasons.

Study 3: Perennial Crops in Eastern Canada

Winter killing is frequently observed and is one of the most common causes of the loss of stands and yield of perennial forage crops in many forage-growing areas of eastern Canada. These crops are directly dependent on the weather and climate for their winter survival, and any change in fall and winter climate or climatic variability may have a significant impact on survival and hence suitability of these crops. The authors of this study[4] analyzed daily weather data for the 1961-90 period from sixty-nine weather stations within agricultural areas of Ontario, Québec, and the Atlantic provinces. Climate change scenarios for two future time periods (2010-39 and 2040-69) were based on a 2001-year "transient" simulation by the first-generation coupled Canadian General Circulation Model (CGCM1) that included the effects of aerosols (Boer et al. 2000).

Following is a brief summary of the highlights of this study. It should be noted, however, that spatial differences in climate meant that results could differ within the region. (For complete details of the methods used in this work and some of the regional variations in the results, see Bélanger et al. 2001, 2002.)

In general, it is expected that temperatures during the fall-hardening phase of perennial forages will be milder in the future in all regions of eastern Canada to varying degrees, and that this will result in lower levels of cold-hardiness as the crops enter into the winter period than under the current climate. Changes in moisture conditions during the fall-hardening period are expected to be relatively small, however, and are considered as unlikely to have significant effects on fall hardening.

In most regions of eastern Canada, the threat of exposure to subfreezing temperatures in the absence of snow cover during winter is likely to increase, resulting in increased risk of winter damage. Moreover, greater exposure to freeze/thaw cycles and to periods of warm temperature during winter is likely to result in loss of plant hardiness and to increase the risk of ice damage to the root system of perennial forage crops. This effect was expected to be further exacerbated by a shift in winter precipitation from snow to rain, leading to greater ice encasement and smothering.

The net impact of changes in the agroclimatic indices investigated in this study indicates increased risk of winter damage, particularly to forage species that are most susceptible to harsh winter conditions, such as alfalfa and orchardgrass. Modification of agronomic practices and cultivars in the future may, however, be instrumental in limiting the negative effects of climate change on winter survival.

The three studies presented here use climate change scenarios to estimate alterations in specific agro-climatic properties and project their potential impacts on specific types of crop production in two Canadian farming regions. Although the analysis does not include possible effects from alterations in weeds, insects, and diseases under a changed climate, the results are useful for providing the "big picture" view of expected trends. The information represents what might be possible in Atlantic and central Canada and therefore offers a basis for considering future risks and opportunities.

Notes

1   Details for this study are available in Bootsma et al. 2005a, 2005b.
2   See http://www.cics.uvic.ca/scenarios/index.cgi?Other_Data.
3   Details for this study are available in Bootsma et al. 2004.
4   Details for this study are available in Bélanger et al. 2001, 2002.

6

Climate Change Impacts on Agriculture in the Prairies

David Sauchyn

Prairie agriculture is exposed to substantial climate variability and extreme weather. The production systems, agro-ecosystems, and soil and water resources that support agriculture are also highly sensitive to climate. This chapter provides an example of the impact-based approach described in Chapter 2 of this volume. The material presented here considers the potential and net impacts from climate change by examining the exposure of Prairie agriculture to risks and the related climate sensitivity of the soil, water, and ecological resources that enable crop and livestock production. Adaptive capacity in the agricultural sector is consistently described as high; the key component of this chapter is therefore the discussion of net or residual impacts, that is, the degree of climatic variability that could exceed the coping capacity of Prairie farmers above and beyond adaptation in anticipation of climate change.

Prairie Agriculture

More than 80 percent of Canada's agricultural land is in the Prairie provinces. With the exception of early settlements on the rivers of southern Manitoba, Prairie agriculture was first introduced and exposed to the dry and variable climate of the Canadian interior in the late nineteenth and early twentieth centuries. The Euro-Canadian agrarian settlement of the prairies was largely preceded and enabled by the Dominion Land Survey and the building of the transcontinental railway. Establishing agriculture in the western interior was a national policy priority, with the Canadian Pacific Railway (CPR) as a significant supporter. A network of agriculture research stations, created to develop dryland farming practices, were among the first government institutions in western Canada. Thus government and industry invested heavily in the region to dispel the notion that the climate was unfavourable for farming or, as John Palliser declared to the Royal Geographical Society in London in 1859, that a large area was "comparatively useless." In 1881, the influential British journal *Truth* stated: "The Canadian Pacific Railway, if it is ever finished, will run through a country about as forbidding as any on earth." The government of Canada had a different perspective and, with the CPR, sponsored favourable surveys such as one by John Macoun, who concluded that the Prairies were "well suited to agriculture" (Wilson and Tyrchniewicz 1995).

Prairie agriculture is inherently sensitive to climate. A normal growing season has sufficient heat and moisture to produce cereal and oilseed crops and forage throughout the Prairie Ecozone, but "normal" seasons are rare. This region has a mid-latitude continental climate, a regime that is among the earth's most variable climates and that is forecast to change in this century to a greater extent than in the rest of southern Canada. Climate varies among farms according to topography, for example, with the collection of cold air in hollows and valley bottoms and the warmer, drier climate on south- and west-facing slopes.

Because agricultural programs are generally implemented according to the varying productivity of farmland, the suitability of soil landscapes for crop production has been classified beginning in the 1960s with the Canada Land Inventory (CLI). The CLI was used in the 1990s to define marginal land under the Permanent Cover Program. Whereas CLI ratings were based mainly on soil factors, the Land Suitability Rating System (LSRS) includes climate factors influencing the production of spring-seeded small grains (Pettapiece 1995). Under the LSRS, the climatic rating of land reflects the variable that is most limiting.

The moisture index, precipitation minus potential evapotranspiration (P-PE), ranges from no (–150 millimetres) to severe (–500 millimetres) limitation. The temperature index, effective growing degree days (EGDD), is derived from growing season length, growing degree days, and daylength. The growing season is defined as beginning ten days after the average date of mean daily temperature of 5°C and ending on the average date of first frost after July 15. Growing degree days are adjusted to account for longer daylengths with increasing latitude. A few modifying factors are recognized as lowering climatic suitability. These regional factors include excess spring and fall moisture and fall frost, which shorten the growing season or interfere with harvest.

The LSRS climatic variables are listed in Table 6.1 next to a comprehensive list of climatic variables that define agricultural productivity in the Prairie provinces. This list of all relevant variables was compiled from information in the Agroclimatic Atlas of Alberta (Chetner and the Agroclimatic Atlas Working Group 2003) and from a survey to assess information needs for seasonal climate prediction conducted by the Prairie Farm Rehabilitation Association (PFRA) and the Canadian Institute for Climate Studies (CICS) prior to workshops in Winnipeg, Regina, and Calgary (Stewart and O'Brien 2001).

Table 6.1

Climatic variables that influence Prairie agriculture

Climatic parameter	LSRS variables[a]	CGCM2 variables[b]
Temperature/heat		
Daily mean		Mean daily
Daily maximum (°C) (T > 30°C)		Maximum daily
Daily minimum (T < 5°C and T < 0°C)		Minimum daily
Growing degree days	Growing degree days (EGGD[c])	
Frost-free period		
Growing season length	Growing season length (EGGD)	
Potential evapotranspiration	Potential evapotranspiration (moisture index [P-PE])	Evaporation (mm/day)
Precipitation/soil moisture		
Annual precipitation	Annual precipitation	Precipitation (mm/day)
Inter-annual variability	Excess spring and fall moisture index [P-PE]	
Growing season precipitation		
Overwinter precipitation/ spring soil moisture		Snow water content (kg/m²)
Summer soil moisture		Soil moisture (capacity fraction)
Timing of summer rain		
Other		
Extreme weather events		
Duration of bright sunshine (hours)	Daylength (EGGD)	Incident solar flux at surface (W/m²), total cloud (fraction)
Mean wind speed		Mean wind speed (m/s)
Wind gusts		
Prevailing wind direction		

Source: Chetner and the Agroclimatic Atlas Working Group (2003).
Notes:
a Variables used in the evaluation of land suitability (Land Suitability Rating System [LSRS]).
b Output from the CGCM2 (Canadian Centre for Climate Modelling and Analysis, http://www.cccma.bc.ec.gc.ca/). Whether measured or modelled, climatic variables generally have standard definitions. For example, surface temperatures are recorded at 2 m height and wind speed is recorded at 10 m height. More complex variables, such as evaporation and soil moisture, can be measured or modelled using various methods. Therefore, beyond basic temperature and precipitation scenarios, future climatic conditions are usually computed from these parameters rather than being based on the model results.
c Effective growing degree days – an index of growing degree days adjusted for daylength (Pettapiece 1995).

Length of the growing season is a consideration in the northern reaches of Prairie agriculture. Otherwise, a shortage of water is by far the dominant limitation on crop and livestock production for the area. The western interior is the major region of Canada where periodic deficits of soil and surface water define the ecosystems and soil landscapes and have determined most of the adaptation in the agricultural sector. Summer moisture deficits are characteristic of the climate. During droughts, the water demands of plants, ecosystems, and/or human activities are not met for a season or longer (Wilhite 2005). The climate sensitivity of ecosystems, soil landscapes, and human activities is amplified by droughts; wet conditions act as a buffer. There is also a large geographic variation in sensitivity to drought according to type, reliability, and accessibility of water supplies. Most of the runoff in the western Prairies is from snow and glacier melt in the Rocky Mountains. This water enables irrigation in southern Alberta and parts of Saskatchewan, and supplies all the cities in these two provinces. In terms of land area, however, most Prairie agriculture is dryland farming dependent on precipitation and local runoff. The driest (semi-arid) ecoregion, the mixed grassland, has the highest frequency of drought. In some years, such as 1961 and 1988, water deficits affected most of the Prairie agricultural zone.

Past and Current Climate Risks for Prairie Agriculture

The Euro-Canadian history of the Prairies is punctuated by the impacts of drought. The drought of the 1930s was the most devastating, largely because soil and water management had yet to be adjusted for drought and because the Prairies had been settled and farmed almost uniformly rather than according to suitability for crop production. These maladaptations have been largely reconciled, initially with almost total depopulation of the driest areas.

In contrast to the 1930s, the recent droughts of 2000-03, the most severe on record, had different, largely economic consequences. Throughout western Saskatchewan and central and northern Alberta, most climate stations recorded the least precipitation for any single year in 2001 and for any three-year period during 2001-03. Wheaton and colleagues (2005) estimated that in 2001, producers in Alberta and Saskatchewan lost $413 million and $925 million, respectively, in terms of the value of lost crop production. In 2002, these losses were $1.33 billion and $1.49 billion, respectively; farm income was zero in Alberta and negative in Saskatchewan. As the impacts of drought in 2001-02 rippled through the economies of the two provinces, the loss in GDP was about $4.5 billion (Wheaton et al. 2005, 22).

As these data for 2000-01 suggest, the impacts of drought are cumulative as the water deficit grows. The economic impact was much greater in 2002 even though 2001 had less precipitation. Whereas costs are assigned to fiscal or calendar years, the duration of drought is measured in consecutive seasons and years. As drought persists, soil and surface water deficits grow with increasingly serious consequences. Thus duration is probably the key determinant of drought impacts. Nemanishen (1998) explained the challenge of coping with consecutive-year droughts:

> Modern farming technologies and practices now enable farmers to cope with single-year droughts. Most of the light lands in the drought prone areas are now either community pastures or seeded to permanent cover. Yet even with modern technologies, the current wheatlands are not able to yield sufficient returns to justify cropping in the second drought year. During the consecutive-year droughts, the precipitation deficit accumulates and leads to the depletion of the soil moisture in the root zone to a depth of a metre or more ... the impact of the accumulated deficit can bring the grainbelt to the verge of desertification, a process which actually occurred in the 1930's. There is no technology, apart from irrigation, which can sustain either cereal grain or hay production during extended drought periods in the Palliser Triangle.

The historic droughts with the greatest impacts in terms of social and financial consequences occurred during 1939-41 and 2000-03; however, droughts of this magnitude are common in records of the pre-settlement climate of the western interior (Sauchyn et al. 2002, 2003).

Tree rings from the margins of the Prairie Ecozone are especially good proxies of drought because they provide annual data and, in this dry climate, tree growth is sensitive to precipitation and soil moisture. Figure 6.1 is a plot of new tree-ring data (residual ring-width chronology) for the period 1610-2004, from Douglas-fir (*Pseudotsuga menziesii*) from the Wildcat Hills near Calgary, Alberta. A negative departure from the mean ring width represents a dry year. The dark grey shading marks periods of more than ten years without consecutive wet years, that is, only single years with wider than normal tree rings. The lighter shading marks three episodes of seven to nine years of sustained drought. This tree-ring record reveals that prolonged drought is more common than the instrumental weather records indicate. The period 1929-41 was the one long period in the twentieth century that lacked consecutive wet years. Droughts of similar or longer duration occurred four times between 1791 and 1872, shortly before agriculture was introduced to the western plains. This includes a drought in the 1790s when the North Saskatchewan River at Fort Edmonton was too low to enable the transport of furs by canoe, and the droughts of the mid-nineteenth century when John Palliser declared a large area "forever comparatively useless" (Sauchyn et al. 2002, 2003). Sustained drought from 1891 to 1898, preceded by four very dry years in the 1880s, delayed settlement by thwarting early attempts to farm in this region (Jones 1987).

Figure 6.1

Douglas-fir tree ring data for Wildcat Hills, Alberta, 1610-2004

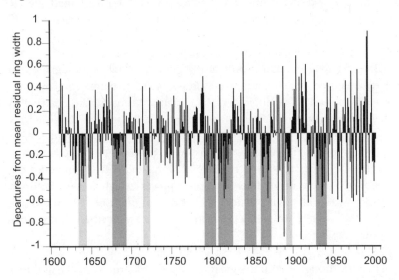

Future Exposure and Sensitivities for Prairie Agriculture

Scenarios derived from global climate models (GCMs) are considered the most reliable forecasts of future climate. The current generation of GCMs are coupled ocean/atmosphere models with greenhouse gas (GHG) forcing estimated for a series of social and economic scenarios, as described in the Special Report on Emission Scenarios (SRES) commissioned by the Intergovernmental Panel on Climate Change (IPCC). The climate change data in Figure 6.2 are derived from CGCM2 B21 (that is, version 2 of the Canadian General Circulation Model with GHG forcing specified by SRES scenario B21). This emission scenario emphasizes global solutions to environmental and social sustainability and therefore produces relatively conservative temperature trends. Figure 6.2 shows mean annual and spring temperatures for 1961-90 and the 2020s (2010-39), 2050s (2040-69), and 2080s (2070-99). These data are averages for the ten GCM cells that comprise the Prairie agricultural zone. They illustrate the significant warming that is expected for the Canadian interior, even using the optimistic B21 emission scenario.

Figure 6.3 is a scatterplot of the change in temperature and precipitation from 1961-90 to the 2050s as forecast by twenty-five GCM experiments for the GCM grid cell that occupies southeastern Saskatchewan. Plots for other grid cells in the Prairie provinces show a similar scatter of scenarios. Most of the model simulations are concentrated in the range of a 2.5-4°C increase

Figure 6.2

Mean annual and spring temperatures in the Prairie agricultural zone, 1961-90, 2010-39, 2040-69, and 2070-99

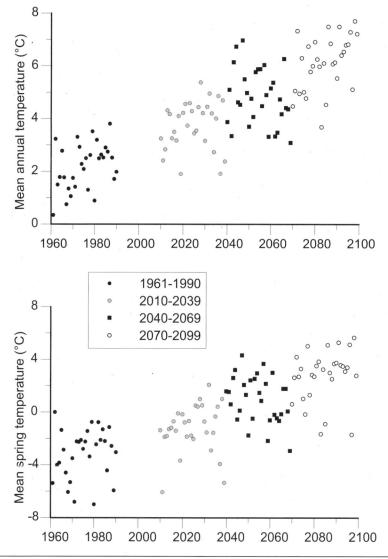

Source: Averaged data from CGCM2 B21.

in temperature and 2-10 percent increase in precipitation. CGCM2 B21 plots are slightly above the median forecast for temperature and below the median forecast for precipitation.

Figure 6.3

Changes in temperature and precipitation in the Prairie agricultural zone, 1961-90 to the 2050s

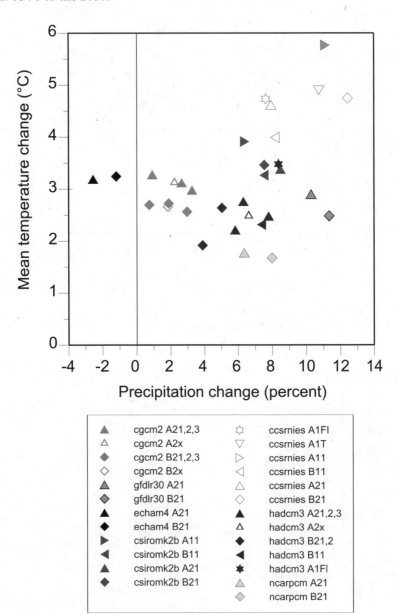

Source: Data from 25 GCM outputs, as noted in legend.

The best-case scenario for Prairie agriculture is based on climate models (e.g., HadCM3; see Figure 6.3) that forecast enhanced precipitation with a moderate increase in temperature and water loss by evapotranspiration (McGinn et al. 2001). Higher productivity would result from the longer growing season, higher temperatures, and concentrations of CO_2, elevated soil moisture, advanced seeding, and accelerated growth and maturation of crops before peak summer aridity. McGinn and colleagues (2001) projected that these conditions would produce a 21-124 percent increase in the yields of canola, corn, and wheat in Alberta.

Other scenarios, derived from the Canadian General Circulation Model, for example, suggest that higher temperatures will force accelerated evapotranspiration and for more hours per year, causing water loss that exceeds a marginal increase in precipitation (Sauchyn et al. 2002). Nyirfa and Harron (2001) used output from CGCM1 and the LSRS to simulate changes in land suitability under the forecasted changes in temperature, precipitation, potential evapotranspiration, and effective growing degree days. The result was a significantly higher moisture limitation for the production of spring-seeded small grain crops over much of the agricultural area. They concluded that the warmer and drier conditions would require adaptation to adjust the distribution of production systems to the new land suitability.

Even though most GCM-based climate change scenarios suggest increasing aridity as water loss from soils and plants exceeds the inputs from a marginal increase in precipitation, a recent analysis of 1951-2000 evaporation data for the Prairie provinces revealed significant decreasing trends in June, July, October, and annual evaporation for 30-, 40-, and 50-year time periods (Hesch and Burn 2005). Increasing evaporation was recorded only for September over the 30-year period, and for April at the longer record length of 50 years. Most of the increase was in the more northern region, while the southern regions showed decreasing trends. The authors did not attempt to explain the decreasing trends, but other studies (e.g., Roderick and Farquhar 2002; Loaiciga et al. 1996) point to the increased humidity and cloudiness forecast by GCMs and observed during the past several decades. Furthermore, models and measurements show that much of the warming reflects the rise of minimum temperatures, that is, at night and in the spring. This could account for the observation by Hesch and Burn (2005) of increasing evaporation in April.

Potential changes in crop yield are the most researched climate impact in the agricultural sector (e.g., McGinn et al. 1999; Thomson et al. 2005). Simulations of the effects of changes in temperature, precipitation, and CO_2 concentrations produce a wide range of crop yield estimates depending on the climate scenario and the sensitivity of the model to the soil water balance and to changes in plant water-use efficiency. These assessments of direct climate impacts on crop productivity have limited application to adaptation planning, because crop yield projections are particularly sensitive to the forecasting of precipitation changes and the parameter that GCMs simulate with the least certainty. They are also based on annual and seasonal climatic conditions, whereas crop yields are strongly influenced by local weather and soil factors, and agriculture in general is most vulnerable to climatic variability and extremes. Furthermore, much of the increased productivity is attributable to the positive effects of higher concentrations of CO_2 in terms of fertilizing crops, reducing transpiration, and improving water-use efficiency (Thomson et al. 2005). There are important temperature and CO_2 thresholds, however, where crop yields level off and potentially decline as water and nutrients become limiting factors.

Future crop yields could also depend on the changing effectiveness of herbicides and pesticides with global warming. Increased efficacy of herbicides at higher levels of CO_2 is generally assumed, but research in Saskatchewan (Archambault et al. 2001) showed a varied response and indicated that the interactive effects of increased CO_2 and temperature caused a decrease in herbicide efficacy. At current rates of CO_2 emissions, the changes in herbicide efficacy will not be apparent for about another fifty years. Any potential economic losses due to decreased herbicide efficacy could be totally or partially offset by increases in yields from higher CO_2 concentrations.

A shift to climatic conditions that favour other crops requires obvious adaptation strategies: substitution of crops or cultivars or a change in land use (Smit and Skinner 2002). Thus, both adverse and positive impacts of higher temperature and CO_2 on crop production can be addressed largely by adaptation. The more serious vulnerability is to conditions that fail to support any crops, that is, insufficient soil or surface water either seasonally (drought) or in the longer term (aridity). These two scales of water deficit are linked since aridity is a function of the frequency, severity, and duration of drought.

While the source of and access to water vary considerably across the Prairies, in general agriculture is most dependent on good-quality surface water derived from snowmelt runoff and, to a lesser extent, spring and summer rain. Probably the most revealing and valuable climate impact studies have therefore been those that have examined future water supplies, especially those derived from the Rocky Mountains, since by far the largest user of water in western Canada is the irrigation industry, and expanded irrigation is often cited as an effective adaptation to increased aridity under global warming (de Loë and Moraru 2004). Various studies (Rood and Samuelson 2005) have documented a decline in surface water in recent decades, and especially snowmelt runoff, but the question remains as to whether these recent trends represent natural cycles or the early impacts of global warming.

Thus, it is useful here to summarize the results of two studies that simulated future water yields from the eastern slopes of the Rocky Mountains, given the major implications for Prairie agriculture.

Demuth and Pietroniro (2001) examined the retreat of glaciers in the Rocky Mountains since about 1850 and runoff in the North Saskatchewan River Basin (NSRB) from 1 August to 31 October, when glacier meltwater is the most significant component of streamflow. They estimated that within 30 to 50 years, glacier cover will have shrunk to the minimum extent for the past 10,000 years. As the glacial cover has decreased, so have the downstream flow volumes. Warmer temperatures should cause increased glacier runoff in the short term. Historical streamflow data indicate that this increased flow phase has already passed, and that the river basins of the western Prairies have entered a potentially long-term trend of declining summer flows.

In the South Saskatchewan River Basin (SSRB), there is much less glacier ice, snowmelt accounts for most of the runoff, and the surface water is already almost fully allocated, since most of the irrigated agriculture in western Canada is concentrated in the SSRB. Lapp and colleagues (2005) modelled future snow accumulation and ablation in the Upper Oldman River watershed, in the SSRB, using scenarios of temperature and precipitation derived from CGCM1. Their simulation suggests that climate warming will result in a substantial decrease in snow accumulation in the watershed, with a corresponding significant decline in spring runoff volumes: "The resulting water scarcity would be especially problematic where development has been based on a more abundant water supply. The irrigation industry, as the greatest single consumer of water in the region, will likely come under the greatest scrutiny and stress" (Lapp et al. 2005).

Another proposed adaptation to increased aridity in western Canada, besides expanded irrigation, is the increasing use of groundwater. Currently, wells are the main source of water for rural households and, under drought conditions, groundwater is the alternative to surface water. A lack of knowledge of prairie aquifers limits information on the availability of groundwater and the modelling of changes in quantity and quality with climate change (Thorleifson et al. 2001). Shallow aquifers are sensitive to climatic variability. Deeper groundwater is less responsive to climatic variation, but it tends to be of poor quality, with high concentrations of dissolved minerals.

The studies cited here demonstrate that Prairie agriculture is exposed to large variations in water supply over time and space, reflecting the sensitivity of prairie hydrology to climate change and variability. Farming and ranching are also vulnerable to the sensitivity of soil landscapes, which is largely a function of climate, since soil degradation is preventable except in situations of aridity and severe drought. Rates of erosion are highest in semiarid landscapes, where there is less protection of the soil from wind and rain. Annual crop production accelerates erosion in these landscapes. The semi-arid to subhumid mixed grassland ecoregion of western Canada is at risk of desertification, which is defined as "land degradation in arid, semi arid and dry/sub-humid areas, resulting from various factors, including climatic variations and human impact" (UNEP 1994, 1,334). When Sauchyn and colleagues (2005) derived climate change scenarios from the CGCM2 (emission scenario B2) and applied these to the modelling and mapping of aridity (P/PET) on the Canadian plains, the area of land at risk of desertification increased by about 50 percent between recent conditions (1961-90) and the 2050s. This impact scenario implies that the soil landscapes of semi-arid southeastern Alberta and southwestern Saskatchewan may require further improvements to soil and water management practices to prevent degradation and sustain agriculture.

Net Impacts from Climate Change on Prairie Agriculture

Scholarship and commentary on Prairie agriculture cite the high adaptive capacity of the sector, which has evolved from the many adjustments of practices and policy in response to climatic, social, and institutional crises and change over a relatively short history. Thus, Prairie agriculture is adapted to the historical range of climatic variability (Hill and Vaisey 1995) and to the distribution of soil and water resources.

Since the settlement of the Canadian Prairies in the nineteenth and early twentieth centuries, land use and farming practices have evolved to match the various climates and soil types on the Prairies and have adapted to changing markets, technology, and transportation systems. The abandonment of farms in the Special Areas of Alberta during the early 1920s, and southwestern Saskatchewan in the 1930s, provides evidence of these adjustment processes. More recently, since the 1980s, there has been a reduction in the summerfallow and an expansion of crop varieties, particularly in areas of higher moisture (PFRA 2000, 81).

Current actions and attitudes of producers are further indication of adaptive capacity. For example, when a group of Alberta producers were asked about recent changes to their operations, almost half indicated that it was because of climate change; "the participants also stated they would implement any necessary strategies to adapt to climate change ... it is not about if they will adapt, but when they will adapt" (Stroh Consulting 2005, 3).

This suggests a low vulnerability to further climate change and variability if the resilience, adaptability, and resourcefulness of Prairie producers and institutions are applied to future trends and events. When de Jong and colleagues (1999) included adaptive crop management strategies in a simulation of climate impacts on crop yields, the net impacts were found to be minimal across Canada. They concluded that there remains room for adaptation, such as more efficient irrigation technology and water distribution and storage, and adjustment of planting dates according to the availability of water. Even improved water management and conservation, however, "are likely to prove effective in mitigating the impacts of extreme climate events, such as the 2002 Prairie drought" (CCIAD 2002, 11).

Even though Prairie agriculture is adapted to the range of historical climatic variability of the twentieth century, there are always climate thresholds beyond which activities are not economically viable. Further adaptation is required only if climate change is expected to result in variability that will exceed the historical experience. The climate scenarios and impact scenarios reviewed here suggest that a drought of unprecedented duration is the climate event most likely to exceed the coping capacity of Prairie producers and agricultural institutions.

We argue, therefore, that the greatest climate risk to the future of Prairie agriculture will be the recurrence of drought of longer duration than has occurred since the settlement of the Canadian plains. Both GHG-forced climate change scenarios and proxy records of natural climatic variability suggest that severe droughts are probable in the twenty-first century. The capacity to cope with threats to the sustainability of Prairie agriculture is illustrated by the adjustments made to practices and policy to address the risk of soil degradation.

The "1980s saw not only a new cycle of prairie drought but also a new environmental thrust with enhanced focus on soil conservation and sustainability" (Vaisey et al. 1996, 2). With the adoption of soil conservation practices, particularly reduced tillage, the average number of bare soil days dropped by more than 20 percent in the Prairie provinces between 1981 and 1996, with a 30 percent reduction in the extent of land at risk of wind erosion (McRae et al. 2000). Institutional responses included the soils component of the Canada-Saskatchewan Agriculture Green Plan of 1990, and the National Soil Conservation Program (NSCP) of 1989. In the Prairie provinces, a major component of the NSCP was the Permanent Cover Program (PCP) (Vaisey et al. 1996).

Conclusions

Prairie agriculture has been sustained through the adjustment of land-use and management systems to climatic variability (e.g., drought, early frosts, storms) to take maximum advantage of soil and water resources (Hill and Vaisey 1995). There is, however, a perpetual adjustment to weather and climate, and now the agricultural sector is faced with the prospect of climate change, which may include climatic variability that exceeds the historical experience.

Potentially positive impacts of climate warming on agriculture arise from correlated changes in temperature and CO_2. These are the climatic parameters that are modelled and measured with the greatest certainty. Recent observations of increased mean annual and spring temperatures are consistent with climate change scenarios. The adverse impacts of climate change are mostly related to increasingly variable and declining soil and surface water supplies. These parameters, and precipitation, are measured and modelled with much less certainty.

The agriculture sector in the Prairie provinces has shown considerable resilience by adapting crop and livestock production to a cold subhumid climate that includes a large area with the least precipitation (less than 330 centimetres) in the continental interior, periodic drought, and seasonal and inter-annual variations in temperature and precipitation that are among the largest on earth. Marginal conditions over a large area, Palliser's Triangle, have fuelled a long-standing debate over whether agriculture should have been introduced in the first place. Adaptations in recent decades have included widespread soil and water conservation and a new and evolving policy framework. The success of these strategies is reflected in the impact of the drought of 2000-03. Despite substantial crop losses, the social impacts did not compare with consequences of the less severe droughts of the 1930s.

It may be that, because the level of adaptive capacity is sufficiently great, the net impacts of climate change will be minimal – that is, the producers and governments have the capacity to expand the coping range to encompass the forecasted climate trends. This will not occur, however, without planned proactive adaptation, unlike the historical circumstances of adjustments to policies and practices largely in response to extreme events and especially drought. The most challenging future climate will include droughts that are of longer duration than those of the 1930s, similar to those that occurred frequently just before most of the region was settled by Euro-Canadians. There are few existing strategies, other than government assistance, to sustain agriculture through these most extreme conditions.

Acknowledgments
The tree-ring research was funded by the Natural Sciences and Engineering Research Council and the Canada-Saskatchewan Agriculture Green Plan. Julie Frischke, Jodi Axelson, Laura Pfeifer, and Melissa Ranalli assisted with the collection and analysis of climate and tree-ring data.

7
Agricultural Water Supply in the Okanagan Basin: Using Climate Change Scenarios to Inform Dialogue and Planning Processes

Denise Neilsen, Stewart Cohen, Scott Smith, Grace Frank, Walter Koch, Younes Alila, Wendy Merritt, Mark Barton, and Bill Taylor

This chapter reports on the agricultural component of a study that developed integrated climate change and water resource scenarios for the Canadian component of the Okanagan Basin in British Columbia (Figure 7.1) (Cohen et al. 2004, 2006). The study serves as a basis for understanding some of the challenges facing both suppliers and consumers of water, currently and in the future, particularly with regard to climate change. Rather than developing an all-inclusive model of water resources, in which the connections to climate and management decisions are mathematically expressed, the approach used dialogue with stakeholders to provide information that could complement mathematical models. The focus was on providing information about future climate change, downscaled from scenarios. Such a technique enables the inclusion of issues that may be difficult to model quantitatively. Mathematical models were used strategically to generate information on known environmental indicators and processes related to climate (temperature, precipitation, snowpack, runoff, streamflow, crop water demand). Dialogue was used for those indicators and processes that include a human component (irrigation, land use, ecosystem requirements, and institutional arrangements).

This study incorporates some elements of the context-based approach by considering land use, water demand, institutions, and so on; the research also includes stakeholder interaction. Its assessment of impacts is based upon climate change scenarios, however, as the stakeholders' consideration of adaptations relies on the established climate change impacts.

In the study, scenarios were developed for the effect of climate change on regional average water demand for agriculture in the Okanagan Basin and for inter-annual variability in demand for a large agricultural water purveyor (Neilsen et al. 2004a, 2004b). This chapter includes both sets of scenarios. They are compared to hydrological responses to the same climate data (Merritt and Alila 2004). The chapter also includes some responses

Figure 7.1

The Okanagan Basin, British Columbia

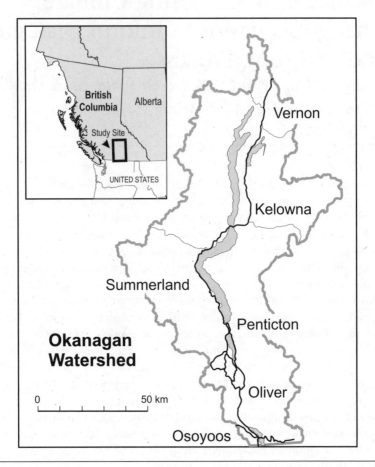

to the challenges associated with inadequate water supply and their implications for future sustainability in the region. Data from a case study of conditions in Summerland/Trout Creek inform the discussion.

The chapter begins with a review of the pressures facing water resources in the Okanagan Basin and of agricultural production in the area. A discussion of the processes behind the development of models for the study area (Summerland/Trout Creek) is presented, followed by results from the case study region. The chapter concludes with details from workshops where dialogue with end users took place, and discusses implications of the study's findings.

Pressure on Water Resources in the Okanagan Basin

Water resources are known to be sensitive to variations in climate, and will likely be influenced by climate change projected from current and future increases in greenhouse gas concentrations in the atmosphere. In western North America, most rivers are dominated by snowmelt runoff, which often comprises 50-80 percent of annual flow. Since the late 1940s, a shift in the timing of snowmelt runoff towards earlier dates has been observed (Stewart et al. 2004). This has been strongly connected with observed increases in spring air temperatures. Projected warming during the twenty-first century could lead to a continuation of this trend, with implications for reservoir management and water users (Stewart et al. 2004). Among the implications is a reduction in flow, as noted in Chapter 6.

The Columbia Basin has been the subject of some detailed case studies (Hamlet and Lettenmaier 1999; Miles et al. 2000; Mote et al. 1999, 2003; Payne et al. 2004), and there has been an initial attempt to bring in trans-boundary perspectives (Cohen et al. 2000). This work suggests that a warmer climate would lead to changes in hydrology, including reduced snowpack and earlier snowmelt peaks, with subsequent implications for regional water supplies and fisheries. The earlier peak would lead to increased flow during winter months and an earlier flood season. Less water would flow during the summer months, when irrigation demand is highest.

The Okanagan sub-basin is subject to some of the potential pressures identified for the whole Columbia River system in response to climate change (Barnett et al. 2004; Mote et al. 2003), particularly the conflict over allocation of water resources for in-stream use by fish and wildlife, consumptive use in agriculture, and development of non-irrigated land to support current, rapid population growth. A number of issues arise out of the intersection of natural constraints on the system, current pressures, and future climate change. In addition to potential changes in water supply and demand for agriculture, these include the great sensitivity of semi-arid areas to small changes in precipitation; challenges to decision makers to integrate land- and water-use planning; incremental effects of peak and low flow timing and water temperatures on fish habitat; economic costs associated with development of new water supply and storage; and control of major lake levels to support conflicting requirements for flood control, recreation, in-stream requirements, tourism, and so on.

Agriculture in the Okanagan Basin

Agriculture accounts for 75 percent of consumptive water use in the Okanagan Basin, where crop production is entirely dependent on irrigation. Currently, the climate of the region supports primarily perennial crops (high-value tree fruits and wine grapes, with the balance in pasture and forage) and a small acreage of annual crops (potatoes, vegetables) planted in suitable microclimates. For high-value horticultural crops, successful production is an issue of quality rather than quantity (yield), as will be pointed out in Chapter 12. Management practices focus on improving quality attributes. In much of North America and throughout the world, such crops are grown under irrigation, as high returns warrant expenditures on expensive infrastructure. Timely availability of water is imperative to economic production of high-value crops, both to assure quality and to protect investment in perennial plant material. In some crops, such as wine grapes, planned water deficits are used to enhance quality attributes (Behboudian et al. 1998; Dry et al. 2001) while conserving water. Consequently, potential limitations and adaptation to the availability of irrigation water under current and future climates are important considerations for agriculture.

Changes in average climate will determine, in the long run, which crop production systems are viable in a region. Extreme climate events present a greater challenge to agriculture and to communities and their ability to cope, however (IPCC 2001). One of the major risks facing Okanagan agriculture is the occurrence and frequency of drought. In production systems that are entirely dependent on irrigation, drought can be defined as the inability to provide an adequate water supply to maintain an economic return. There are two components to drought in this case: high demand and low supply. It might be expected that the most dramatic effects will occur when high demand and low supply are combined. The risks associated with drought are determined by the severity and frequency of occurrence of drought conditions. Forecasting when droughts may occur is not feasible. It is possible, however, to examine the historical record and determine how frequently periods of high water demand and low supply have occurred, and, under climate change scenarios, to determine the likelihood of such combinations in the future.

Model Development for the Study
A GIS (Geographic Information Systems) approach was used to model crop water demand. Current and future climate data were spatially interpolated to a grid and superimposed on a land-use cover to allow crop-specific calculations of water demand.

Climate Change Scenarios
Despite their increasing sophistication, GCMs are still limited in the detail with which they simulate climate and may produce a range of results depending on the model constraints and the set of conditions for which a model experiment is run. One of the major sources of uncertainty in GCMs is the magnitude of future greenhouse gas emissions, and a set of emissions

scenarios has been developed based on global population projections, economic growth, and technological change (Nakicenovic et al. 2000). In order to account for uncertainty associated with GCM output, a range of GCMs and emissions scenarios was used in the current study (Taylor and Barton 2004), as recommended in the IPCC *Third Assessment Report* (2001). These included the Canadian (CGCM2), Australian (CSIROMk2), and British (HadCM3) models with two emissions scenarios (SRES) for each model: A2 – higher rate of emissions growth with regional economic development and a continuously increasing population, and B2 – lower rate of emissions growth with local solutions to environmentally sustainable development and a moderate rate of population increase. Output from GCM experiments is in gridded data sets with a grid size of several hundred kilometres. In constructing regional climate scenarios, it is common practice to take into account data from several surrounding grid cells. This may be misleading, however, if neighbouring grid cells contain very different terrain, as in the case of the Okanagan Basin, where adjacent cells contain either ocean or prairie (E. Barrow, pers. comm.). Consequently, as the Okanagan Basin was contained within a single cell for each GCM, single grid cell data were used.

Downscaling

There has been considerable discussion regarding the scale of climate change scenarios when modelling impacts in agriculture and hydrology (e.g., Mearns et al. 2003; Payne et al. 2004). Giorgi and Hewison (2001) indicate that high-resolution scenarios would probably provide more realistic responses than coarse-scale scenarios, particularly in mountainous regions. At present, there is no suitable regional climate model (RCM) available for southern British Columbia to allow dynamical downscaling from GCM scenarios to a finer spatial scale (~45 km × 45 km). Construction of annual variation in future climate has been attempted through the use of weather generator models (Semenov and Barrow 1997) and statistical downscaling (Wilby et al. 2002), but techniques were not available to extend such model output to the spatially distributed climate data based on multiple climate stations and topography that was used in the current study.

In a preliminary study, Taylor and Barton (2004) indicated that, although the Statistical Downloading Model (SDSM) was somewhat successful in generating scenarios for daily maximum temperature, it was not possible to produce meaningful precipitation scenarios, as very weak correlations were found between surface precipitation and predictor variables. Consequently, statistical downscaling was discounted as a viable option for the hydrology modelling portion of this project (Merritt and Alila 2004). Instead, a much simpler approach, the *delta method* was selected, in which daily or monthly station data are perturbed by future changes in the thirty-year averages of monthly mean, maximum, or minimum temperature and precipitation derived from GCMs.

Despite the many limitations of this method, not the least of which is the inability to account for changes in the variability of climate data over time (Daly et al. 2002), there are tangible benefits, including the availability of a range of GCM and SRES scenario outputs, which satisfies the need to account for uncertainties associated with GCMs (IPCC 2001); relative ease of computation for gridded climate data sets; and comparability with other studies that have used the delta method (Lettenmaier et al. 1999; Morrison et al. 2002). In order to maintain consistency among all components of the study, the delta method of perturbing 1961-90 normals station data with average monthly changes from GCM output was also used to determine climate change scenarios for crop water demand modelling.

Spatial Climate Distribution
In highly complex terrain, temperature and precipitation may vary over short distances, and these differences may not be captured by widely spaced weather stations. A range of statistical techniques has been developed to interpolate weather station data spatially with respect to latitude, longitude, and elevation. One such approach is to combine statistical interpolation techniques with expert information about local climate regimes. PRISM (parameter-elevation regressions on independent slopes) is an expert system that generates gridded data sets by interpolating among weather station data sets, and also includes "facets," which are map elements defined by topography and terrain with known climate anomalies (Daly et al. 1994, 2002). The definition of facets is based on expert knowledge. PRISM data sets for thirty-year normal, monthly precipitation and temperature are available for British Columbia and were adapted for use to determine both current and future temperature scenarios. Observed differences between PRISM estimates and weather station temperatures led Neilsen and colleagues (2001) to conclude that data based on the original 4 km × 4 km grid cell output from PRISM likely underestimate temperatures in crop-use polygons because of the large elevation changes within the 4 km × 4 km grid cells. In consultation with Chris Daly, the 4 km × 4 km gridded data set was rescaled to a 1 km × 1 km grid by calculating local average lapse rates based on existing grid cell temperature and elevation data. Data from the twenty-four nearest neighbour cells were pooled to calculate lapse rates.

Temporal Interpolation of Climate Data
A second transformation of scenario data involved the derivation of daily minimum and maximum temperature values during the growing season from PRISM monthly climate data (Tmax, Tmin). Daily mean temperature

Figure 7.2

Generalized crop coefficient curves for tree fruit and grapes

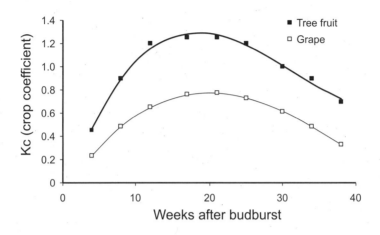

Source: Also published in Neilsen et al. (2006, 927).

estimates were required to calculate growing degree-day accumulations, and daily maximum temperature was required to calculate potential evapotranspiration (PET). Each monthly average was assigned to the middle of the month. For climate change scenarios, monthly data were perturbed by deviations (bias) in monthly maximum and minimum temperatures derived from GCM output (Figure 7.2). In order to ensure a smooth transition between months, a smoothing algorithm was defined, which redistributes the discontinuities between perturbation factors and monthly temperatures throughout the month according to the method of Morrison and colleagues (2002). Details of how smoothing was achieved for both the basin-wide study and the Trout Creek/Municipality of Summerland/Trout Creek case study are given by Neilsen and colleagues (2004a). The end result is a thirty-year climate record that was synthesized for each PRISM grid cell for both 1961-90 and periods centred on the 2020s, 2050s, and 2080s.

Precipitation
Spatial rescaling of 4 km × 4 km PRISM precipitation was not possible. Daily precipitation data for four representative Environment Canada weather stations with long-term records – Vernon Coldstream/CS, Kelowna Airport, Summerland CS, and Oliver STP – were used to estimate precipitation in each of the four climate zones defined in Figure 7.1. Precipitation data were modified by monthly GCM scenario biases as described above.

Land Use

The complex land-use patterns of orchards and vineyards were constructed from several cycles of land-use survey. These are viewed as adequately reflecting current conditions. Within the Okanagan Basin, irrigated croplands comprise only a portion of the total agricultural land base, which also includes large areas of non-irrigated rangeland. A map and related database of irrigated cropland were compiled from a number of sources and incorporated into a GIS using ArcInfo™. Information on vineyards is from Bowen and colleagues (2005), tree fruits from the Okanagan Valley Tree Fruit Authority (1995), other croplands from the British Columbia Ministry of Sustainable Resource Management (2001), and pasture lands from the Canada Land Inventory (1966) updated from current cadastral survey data to eliminate urbanized areas. Location data for vineyards and tree fruits are relatively detailed. In areas dominated by pasture and cropland (i.e., in the North Okanagan), land-use boundaries were "ground-truthed" and redrawn as necessary. The interaction of land use, crop water demand, and licensed water allocations was examined by including digitized boundaries of water purveyor districts within the GIS.

Crop Water Demand

The calculation of crop water demand requires some estimate of evaporative demand determined by weather conditions, and plant response determined by growth stage and water availability. For annual crops, demand is often simulated using the water-use efficiency component of growth models and evapotranspiration (ET) data. For perennial crops, the annual development of the canopy is a major determinant of the response to evaporative demand. In irrigated crops, soil moisture is likely not a limitation to growth, and water demand for irrigation can be estimated by applying a canopy development factor to measured or estimated ET (Doorenbos and Pruitt 1977). This approach was considered to be particularly appropriate for the current study, where the objective was to determine basin water requirements for irrigation.

Limitations in the climate data (only rainfall and temperature data were available from PRISM) meant that simple relationships were required to estimate potential ET. Algorithms to estimate daily potential evapotranspiration (PET) during the growing season (approximately 1 April to 1 Nov) were developed by comparing daily maximum temperature (Tmax) and extraterrestrial radiation (Ra) at the Summerland CDA. Environment Canada weather station against ET estimates from an electronic atmometer (Etgage Company, Loveland CO.). A PET value was calculated for each PRISM cell as:

$$PET = -3.26 + 0.210 \text{ Tmax} + 0.058 \text{Ra} \quad (R^2 = 0.58)$$

The solar energy (MJ m^{-2}) reaching the top of the atmosphere (Ra) was calculated from day of the year and latitude. Estimates from the equation compared well with Penman-Monteith ET$_o$ calculations (Allen et al. 1998) for 1994-98 ($R^2 = 0.8145$), but tended to underestimate ET$_o$ at low evaporative demand and overestimate ET$_o$ at high evaporative demand.

Crop coefficients dependent on canopy development were used to modify PET. Maximum mid-season crop coefficients (based on Penman-Monteith ET$_o$) range from 0.9 to 0.95 for apricot, peach, pear, and plum, and from 0.95 to 1.0 for apple and cherry under clean cultivated conditions (Fereres and Goldhammer 1990). These may be expected to be 20-30 percent higher under a cover crop (Doorenbos and Pruitt 1977). Changes in projected temperature from 1961 to 2100 required a crop coefficient curve that could be adapted to a longer growing season. For the purpose of the current climate change study, seasonal crop coefficient curves for tree fruit and grapes (Figure 7.2) were derived from generalized curves based on those published by Doorenbos and Pruitt (1977). Water use for pasture, forages, and other field crops was not modified by seasonal crop coefficients, and was thus characterized directly by calculated PET values.

A second factor in estimating irrigation demand is the length of the growing season. Budbreak in deciduous trees is related to genetic factors and spring air temperatures, and to the magnitude of minimum winter temperatures. For the current study, we examined the relationships between degree-day accumulations and bloom date for a number of tree species, in a complete and well-recorded set of unpublished data collected for phenological studies between 1937 and 1964 at the Pacific Agri-Food Research Centre in Summerland. A set of equations for all tree fruit species was based on the start of growing degree days base 10°C (GDD10) accumulation and date of apple bloom, and the relationships derived for apple and other species (Neilsen et al. 2004a). The start of the growing season in grapes has similarly been related to the start of GDD10 accumulation (Association of British Columbia Grape Growers 1984). For pasture and perennial forages, the start of the growing season to the start of growing degree days base 5°C (GDD5) accumulation was used as the start of the growing season. The end of the growing season was determined as the end of accumulation of GDD5.

Geographical Information System

PRISM-gridded climate scenario output data were assembled in ArcInfo™ and intersected with the agricultural land-use coverage. This procedure created a database that described climatic conditions, over the year, for each unique land unit (polygon). Daily time-step calculations of crop water demand and summary of annual, monthly, or daily values of PET, growing degree days base 5°C and 10°C and volume of water demand were made in the database for each land-use polygon.

Water Supply

The Land and Water BC water licence database was queried for water licences for irrigation purposes in the Okanagan Basin. There are 1,201 water licensees, including 10 large, 7 medium, and 34 small, publicly owned water purveyors, with 3 large and 1,147 small, private holdings. The three major sources of water are the Okanagan River and lake-system main stem, tributary streams, and groundwater. Approximately 75 percent of irrigation water comes from headwater diversions and high-elevation, in-stream storage basins; the remainder is pumped from lakes, streams, and groundwater (Agrodev Canada 1994). Estimates of water supply derived from the University of British Columbia (UBC) watershed model (Quick 1995) and parameterized for the Okanagan Basin were used for supply/demand comparisons. Details of the analytical methods, assumptions, and limitations of the watershed modelling exercise are given by Merritt and colleagues (2006). There were insufficient data for a whole basin water budget to be constructed, so comparisons were made between modelled basin-wide crop water demand estimates and modelled Okanagan Lake inflow; in a case study of the Trout Creek watershed, estimated supply was compared with crop water demand within the Municipality of Summerland under the range of climate change scenarios described previously.

Okanagan Basin Demand and Supply Scenarios

Demand

Crop water demand scenarios were created for the whole basin. Estimated demand for 1961-90 was around 200 $m^3 \times 10^6$. All models showed an increase in annual crop water demand over the next 100 years, up to a maximum of 324 $m^3 \times 10^6$ (Figure 7.3). The HadCM3 model showed the greatest increases and the CGCM2 model scenarios the least. This is consistent with the range of summer temperatures estimated from the three models, as the crop water demand scenarios are dependent on the effects of temperature on both ET and growing season length.

There was little difference between crop water demand scenarios for CGCM2 and CSIROMk2 under the high emissions scenario (A2). A 12 percent increase above historic (1960-91) crop water demand was projected for the 2020s, and a 24 percent and 40-48 percent increase for the 2050s and 2080s, respectively. For the HadCM3 model, demand was projected to increase by 20 percent in the 2020s, 38 percent in the 2050s, and 61 percent in the 2080s. The last value is likely unrealistic as temperature increases as high as 11°C were projected for midsummer in the 2080s. For all GCMs, modelled crop water demand for the 2020s and 2050s time slices and low emissions scenario (B2) was similar to those for the A2 scenario, but lower by the end of the century. The lack of difference between high and low

Figure 7.3

Estimated average annual crop water demand for the Okanagan Basin under historic conditions and six climate scenarios at three time slices

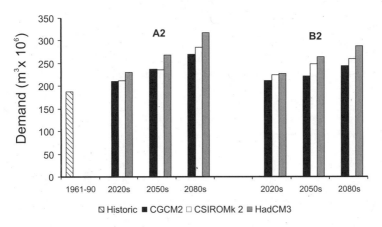

emissions scenarios in the earlier part of the century can be explained by the different emissions scenarios for carbon dioxide and sulphur dioxide. In the near future, temperature increases associated with the two different emissions scenarios will be partially due to current greenhouse gas emissions. The warming expected due to CO_2 emissions is offset by accompanying SO_2 emissions, which are projected to be higher under A2 than under B2 scenarios (Albritton and Filho 2001).

Model output for the length of the growing season indicated an increase from an average of 194 days for 1961-90 to a maximum of 258 by the 2080s (Figure 7.4). Changes in the length of the growing season, and hence in the requirements for water supply, will put added strain on the ability of water purveyors to meet demand, over and above the effects of increased peak evaporative demand mid-season.

Basin Supply
The UBC watershed model was used to estimate Okanagan Lake inflows (Merritt et al. 2006) under current and future climates as an indicator of basin water supply. As a caveat, it should be noted that this is not a quantitative measure of total basin supply and thus cannot be used to calculate a water balance. Current average net annual inflow into Okanagan Lake is estimated to be 450-500 $m^3 \times 10^6$. Model estimates of average total inflow with no restrictions, withdrawals, or surface evaporation were 840.3 $m^3 \times 10^6$ for the 1961-90 period, remaining constant or declining slightly for 2020s scenarios (86-105 percent), and decreasing to between 80 and 93

percent of current inflow for 2050s scenarios and to around 69-84 percent
for 2080s scenarios (Figure 7.5). When compared with estimates of supply
over the same period, it is apparent that, on average, future model scenarios
that projected an increase in demand also led to a decrease in supply.

Figure 7.4

**Estimated change in the length of the growing season for the Okanagan
Basin under historic conditions and six climate scenarios at three time
slices**

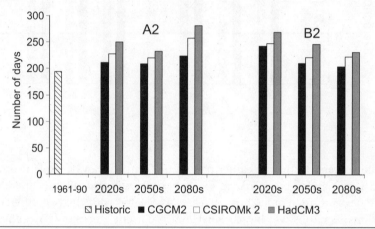

Figure 7.5

**Comparison of estimated demand for irrigation water in the Okanagan
Basin and estimated inflow into Okanagan Lake at three time slices with
three GCMs**

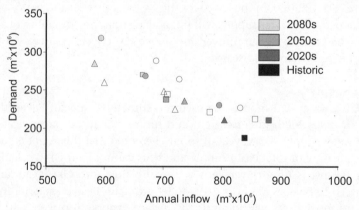

Note: GCMs: CGCM2 (■), CSIROMk2 (▲), and HadCM3(●). Filled symbols are A2 scenarios;
open symbols are B2 scenarios.

The current total licensed allocation for irrigation (323.7 m^3 × 10^6) estimated from the Land and Water BC water licence database is much larger than modelled demand for the 1961-90 period (183 m^3 × 10^6). Under climate change scenarios, the demand for agricultural water use projected over this century ranges from 210 to 317 m^3 × 10^6 (Figure 7.5). The trend towards reductions in supply projected over the same time period suggests that increased allocation of water for agriculture will be unlikely in the future.

Summerland/Trout Creek Case Study

Limitations in the water supply system, in terms of both water withdrawal rights and the ability to store water, likely determine community vulnerability to the effects of drought, particularly if agriculture is the dominant economic activity. Jones (2000) examined the risk posed by climate change to agricultural viability in an irrigated production system in south Australia. Risk was defined as the probability of exceeding known water supply thresholds. In the Trout Creek study, risk thresholds associated with licensed limits on demand and hydrological limits on supply are defined and compared with projected demand and supply.

Future Water Demand

Peak measured consumption since 1990 (10 m^3 × 10^6) occurred in 2003, which was a year of severe water shortage in the Trout Creek watershed. This value is used as a demand threshold against which model output for crop water demand is compared (Figure 7.6[a]). Demand increased over time compared with historical demand (1961-90) for all scenarios. Based on observed climate data, demand for 1991-2003 was as high as that projected for CGCM2 and CSIROMk2 scenarios for the 2020s. The number of years when the demand threshold might be exceeded increased over time, to around 18 out of 30 years by the 2080s for CGCM2-A2 s, 28 out of 30 for CSIROMk2-A2, and 30 out of 30 for HadCM3-A2 scenarios. For the lower emissions scenarios (B2), differences in response compared with the high emissions scenarios (A2) became evident only by the end of the century.

Future Water Supply

Outputs from the UBC watershed model for Trout Creek (Merritt and Alila 2004) were used to derive water supply estimates. From historical patterns, average unrestricted flow in Trout Creek had previously been estimated as 2.65 m^3/sec (Northwest Hydraulic Consultants 2001). Based on the UBC watershed model, modelled unrestricted flow at the mouth of Trout Creek averaged 2.83 m^3/sec for the period 1961-90. For the 2020s, similar unrestricted flows were projected in CGCM2 scenarios and the HadCM3-A2 scenario. Projected flow rates were slightly higher in the HadCM3-B2 scenario

Figure 7.6

Magnitude and variability of modelled crop water demand (top row) and total annual flow (bottom row) in Trout Creek in response to climate change scenarios

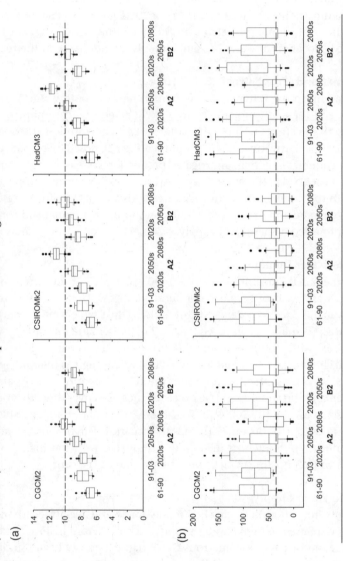

Note: The dashed line in top row (a) is a maximum demand threshold imposed to protect in-stream flow requirements. The dashed line in the bottom row (b) is defined drought for Trout Creek.

Source: Associated Engineering (1997).

and were lower in CSIROMk2 scenarios (Neilsen et al. 2004b). Reductions in flow were projected for the 2050s and 2080s in all model scenarios.

A drought threshold of 30.3 m^3 × 10^6 (36 percent of average annual flow) has been proposed for Trout Creek (Associated Engineering 1997). Total annual unrestricted flow volumes were calculated for each of the GCM scenarios and compared with modelled unrestricted flows (1961-90) (Figure 7.6[b]). Between 1961 and 1990, modelled annual flow was lower than the drought threshold in three years. Drought frequency for high emissions (A2) scenarios increased to 5-6 out of 30 years by the 2020s; 8-14 out of 30 years, depending on GCM, by the 2050s; and 13-22 out of 30 years by the 2080s. The driest A2 scenarios were found for the CSIROMk2 model (Figure 7.6[b]). Low emissions (B2) scenarios for the CGCM2 and HadCM3 models produced fewer droughts than A2 scenarios. In contrast, the CSIROMk2-B2 scenarios produced more drought years in the 2020s and drought frequencies similar to A2 scenarios in the 2050s and 2080s.

Risks to agriculture are highest in years where low supply and high demand coincide. There were no high-risk years between 1961 and 1990. The frequency of high-risk years increased under high emissions scenarios for the HadCM3 model in the 2050s (1 year in 6) and for all GCMs by the 2080s (1 year in 4 to 1 year in 2). There were fewer high-risk years under low emissions scenarios.

These supply/demand scenarios indicate that existing infrastructure will be inadequate to meet future needs. Conservation measures alone, which may potentially save 30-40 percent of water used, will be insufficient to meet demands, and future viability will depend on effective water storage. In addition, there is likely to be a greater dependence on stored water to supply requirements. In years of high demand and early snowmelt, the switch from in-stream flow supply to stored water will occur earlier in the growing season, possibly resulting in water shortages. Early use of stored water (before 1 July) occurred six times between 1974 and 2003, five of them since 1992.

Adaptation Options

A number of potential adaptation options affecting either water supply or demand in the Okanagan Basin were explored and associated costs were estimated (McNeil 2004). Demand-side options include water conservation alternatives such as irrigation scheduling, public education, metering, and adoption of efficient micro irrigation technologies. Supply-side options include increasing upstream storage and switching to the mainstream lakes or rivers as a supply source, thereby relying on the large storage capacity of Okanagan Lake.

The costs of both demand- and supply-side options will vary greatly depending on various features of the individual water supply systems and the

Table 7.1

Estimated costs of a range of adaptation options

Option	Cost per acre foot ($)	Potential water saved or supplied
Irrigation scheduling – large holdings	500	10%
Lowest-cost storage	600	Limited
Lowest-cost lake pumping	648	0-100%
Public education – large and medium communities	835	10%
Irrigation scheduling – small holdings	835	10%
Medium-cost storage	1,000	Limited
Low-cost lake pumping (no balancing reservoir)	1,160	0-100%
Trickle irrigation in high-demand areas	1,500	30%
Average leak detection	1,567	10-15%
Trickle irrigation in medium-demand areas	1,666	30%
Lowest-cost domestic water metering	1,882	30%
Medium-cost lake pumping	2,200	0-100%
Domestic water metering, small communities	2,300	30%
Higher-cost lake pumping	2,700	0-100%
Higher-cost metering	2,700	20%
Highest-cost metering	3,400	20%

Source: McNeil (2004).

type of demands served (Table 7.1). In summary, there is no single least-cost adaptation option for all water systems, since costs will vary significantly from system to system. The lowest-cost option in one area may turn out to be a higher-cost option in other areas. Other factors, such as water quality and treatment options, will also enter into the decision. Often, a combination of options will be necessary in order to achieve full insurance against future water shortages and demand increases.

For water purveyors, it appears that systems that are already near capacity will have to consider costs of at least $1,000 per acre-foot to conserve or develop supplies of water to adapt to climate change. If projections indicate that large amounts of water must be conserved or supplied, then probably $2,000 per acre-foot would be a reasonable figure to consider in future budgets. Site-specific engineering studies would have to follow to obtain more accurate figures.

For individual farmers, there are a number of potential adaptation options. Planting either low-water-use crops or short-season annual crops would be one method of coping with chronic water shortages. This has the disadvantage of reducing market flexibility, however, and may not always be feasible as the current distribution of agricultural land use is determined by microclimate suitability and potential economic returns. It is tempting, for example, to suggest that crops with apparent intrinsic low value and high water requirements, such as pasture/forage, be substituted with high-value wine grapes with low water requirements. However, the value of pasture/ forage is much higher when viewed as a necessary component in a successful dairy or beef operation, and it is likely that the activity is being carried out in terrain that is unsuitable for relatively tender wine grapes and tree fruits. An obvious example of this is the location of pasture and annual crops in valley bottoms, which suffer from late spring frosts due to cold air drainage and pooling. In addition, market conditions may not support a massive expansion of the wine industry, which occupies only a small fraction of valley agricultural land (Figure 7.7).

A second option for farmers is to adopt water conservation methods, including scheduling irrigation to meet evaporative demand, conservative micro irrigation systems, mulching, and amendments to improve soil water-holding capacity and reduce evaporation and planned deficit irrigation (Neilsen et al. 2002, 2003). Each of these may improve annual water-use efficiency compared with unplanned irrigation practice, but there will still be the need to supply water at current allotted flow rates to meet maximum evapotranspiration during midsummer. Thus water conservation cannot be demanded without a thorough understanding of plant requirements and physiology. Some of these adaptations have already been made in that a number of growers employ a range of conservation practices. With funding from a range of sources, fruit growers have established a network of weather

Figure 7.7

Distribution of irrigated agricultural land use in the Okanagan Basin

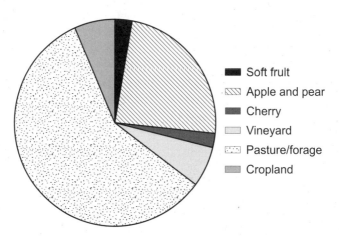

- ■ Soft fruit
- ▨ Apple and pear
- ■ Cherry
- ▢ Vineyard
- ▨ Pasture/forage
- ▨ Cropland

stations to supplement data from Environment Canada stations, and the data are freely available from FarmWest.com with an irrigation scheduling tool. Grape growers have also installed weather stations, and many practice deficit irrigation techniques.

Adaptation to Previous Water Management Issues

Four case studies of water management practices at the local authority level were undertaken to examine how adaptation occurs (Shepherd et al. 2006). Each considers the adoption of a different approach to improving efficiency of water use, including: domestic metering in Kelowna, irrigation metering in the South East Kelowna Irrigation District (SEKID), wastewater reclamation in Vernon, and amalgamation of individual water utilities in Greater Vernon. The last three represent "early adopters" in the region. The four cases demonstrate that adaptation is not a linear, clear-cut process, as action was pursued only when a perceived coping threshold was fast approaching. Response was influenced by option availability, local and provincial agendas and values (including financial incentives), and previous proven effectiveness. Decision and implementation processes were less confrontational when user groups were well informed. Water "savings" were achieved in both metering examples. Kelowna achieved the preset 20 percent reduction target of its single-family metering project. Yearly water allotments in the SEKID were reduced by 10 percent. Although water reclamation was initially implemented as a water treatment strategy (and an effective one), it is now considered as a potential water reuse, and therefore efficiency, strategy.

Lessons learned from these examples are important in answering the question of how to adapt to climate change in the Okanagan. The fear of change and the challenge of transition are important factors affecting the adoption of new policies and practices. Many factors either exacerbate the difficulty of change or smooth out the process; public perceptions and differing political agendas are two key issues. Agricultural users have a strong sense of ownership over water resources due to historical dominance in the region and because storage and distribution systems were originally developed for irrigation. In addition, low water rates are expected, as this has been the norm for many years. There is a divergent relationship among local, regional, and provincial authorities, which may hinder a unified approach in the region, although developments that have occurred since this study was undertaken (see "Conclusions" below) suggest that an integrated approach to water management may be emerging. Finally, any adaptation option to achieve efficiencies may simply allow for further development in the region, unless there is a long-term strategy to conserve the resource for future supply shortfalls due to climate change. This will not reduce local vulnerability to climate change impacts.

Dialogue

Research results were communicated to local water managers, regional planners, producer groups, and water users, in a series of workshops aimed at understanding regional perspectives on how adaptation options might be implemented at the community and basin scales (Tansey and Langsdale 2004). Two locations were selected for community adaptation workshops. Oliver is a small agricultural community in the south Okanagan, and the Trepanier Landscape Unit, including Peachland and unincorporated areas around Westbank, is a rapidly developing area to the west of Kelowna. Participants in the workshops were presented with a range of technically viable supply- and demand-side water management options and were asked to evaluate them according to eight questions that addressed three broad themes: the social acceptability of the options, their current legal acceptability, and political/jurisdictional concerns.

The results of these workshops revealed the complex political landscape that overlay the physical landscape of the region. Historical commitments to users in the agricultural community shape the current allocation of water resources and strongly influence the acceptability of adaptation options. In particular, agricultural users were concerned that current water savings would be reallocated for urban development and not "banked" for future shortfalls. Education and conservation interventions were considered useful in both communities. With respect to groundwater utilization, many participants pointed out that while it may represent a viable alternative source in some areas, extraction is currently unregulated. Increasing drought pressure on traditional water sources may therefore result in largely unregulated groundwater withdrawal.

A third workshop focused on the topic of implementing an adaptation portfolio at the basin scale. The discussion centred on changes that would affect, or be implemented in, the entire region. Because the scale of discussion was broader in geographic area, the adaptation measures discussed were also broader. General supply-side and demand-side approaches were discussed rather than site-specific strategies. There was also a greater emphasis on governance structures that could implement and orchestrate change on this scale. Dialogue at this scale was more strategic. Participants expressed support for expanding the role of the Okanagan Basin Water Board (OBWB) and Okanagan Mainline Municipal Association (OMMA) in regional water quantity management. There was support for basin-wide management of various measures, including increased use of Okanagan Lake and groundwater sources, a coordinated "water smart" program for residential users, and various measures that could be regionally coordinated for agricultural users, such as irrigation scheduling.

One recurring theme at the regional scale was the need for support from the local level, and the need to encourage a sense of "belonging to the basin." Participants also expressed the need for better integration of water issues with local development and planning.

Conclusions

Water supply for agriculture is increasingly important. Irrigated agriculture is often more reliable and efficient than its rain-fed counterpart (Postel 2000), and recent droughts (2000-03) across Canada have increased the demand for irrigation not only in semi-arid regions of the west but also in those areas subject to sporadic water shortages in eastern and central Canada and coastal British Columbia. The Okanagan Basin provides an interesting microcosm for studying the stresses associated with supporting agriculture in an environment where the competition for water and land is already high and is sensitive to climate variability and change.

Currently, the issue of water supply is very much in the public eye. The Okanagan Basin Water Board, an umbrella organization representing the three regional districts (North Okanagan, Central Okanagan, Okanagan Similkameen), recently reconfigured its activities to include a greater focus on basin-wide water quantity in addition to its previous commitments to water quality and eradication of Eurasian milfoil. In addition, the provincial government (through Land and Water BC) has committed to a major modelling effort to develop a basin water budget, in order to determine what water, if any, is still available for licence allocation. The broader business community has made water supply, planning, and agriculture flagship issues in an endeavour to promote sustainable development in the region.[1]

For the farming community, heightened awareness has resulted in increased participation in issues of development, community planning, and withdrawal of land from the Agricultural Land Reserve. The publication of our study (Cohen et al. 2004) has proved timely in helping inform the debate. The combination of modelled scenarios for climate and water supply and demand, in conjunction with an analysis of institutional responsiveness, economic costing of adaptation options, and dialogue sessions with water purveyors, planners, and users, contributed to the usefulness of this study. It would have been greatly reduced in effectiveness if any of the components had been eliminated. As such, it is a successful example of an interdisciplinary approach to climate change impact and adaptation science.

Note
1 For more information, go to http://www.okanaganpartnership.ca.

Part 3:
Context-Based Studies

Part 3 presents examples of studies that provide insights into the context within which adaptations to climate occur in agriculture. This perspective includes a wide range of stressors besides climate, given that producers rarely deal with risks (from climate or other stresses) as isolated phenomena. The highly integrated nature of farming systems means that climate and weather risks are closely linked to decisions regarding yields, input costs, prices, management systems, and environmental factors; they vary according to farm types and locations.

Each of the four chapters in Part 3 addresses an aspect of the context for research on climate change adaptation in agriculture. In Chapter 8, Ben Bradshaw emphasizes the importance of including major socio-economic factors external to the farming system, and cautions researchers and policy makers not to ignore past responses and risk management behaviour when assessing the capacity of agricultural operations to adapt to climate change impacts. In Chapter 9, Henry Venema illustrates many of these ideas with details from Prairie agriculture, while Harry Diaz and David Gauthier focus in Chapter 10 on a specific region of the Prairies (the South Saskatchewan River Basin) to assess the role of institutional capacity in adapting to climate change in agriculture. In Chapter 11, Christopher Bryant, Bhawan Singh, and Pierre André preface their analysis of climate-related risk management in Québec agriculture with a review of how climate change research has developed in that province. Their examination of farmers' perceptions of climate change connects with the impact-based approach (Part 2) through its use of climate scenarios as the starting point, and connects with the process-based approach (Part 4) through its use of information directly from farmers.

8

Climate Change Adaptation in a Wider Context: Conceptualizing Multiple Risks in Primary Agriculture

Ben Bradshaw

This chapter embarks from the simple fact that climate represents only one of many sources of risk (or opportunity) to which Canadian farmers are exposed and respond. Events such as commodity market downturns, changes to government support programs, fluctuations in currency and interest rates, or the loss of export markets due to consumer health concerns may present significant risks to producers at certain times.[1] While this reality has been widely recognized in the literature on climate change impacts and adaptation in agriculture (see, for example, Brklacich et al. 2000; Bryant et al. 2000; Chiotti and Johnston 1995; Eakin 2000; Easterling 1996; Kandlikar and Risbey 2000; O'Brien and Leichenko 2000; Schneider et al. 2000; Smit et al. 1996; Timmerman 1989), the complexity of assessing farmers' exposure and response to climatic *and* non-climatic signals has hindered research that seeks to explicitly incorporate the wider context.[2]

The first step in overcoming this complexity is to conceptualize better the place of climate risks relative to other risks in primary agriculture, the interaction of climate risks with other risks, and the means by which producers adapt to these multiple risks. This is the primary aim of this chapter. The next section makes use of the agricultural systems model and the vulnerability approach to reflect farm-level exposure and adaptation to multiple risks. It is followed by a section identifying a series of trends that arguably comprise a historical trajectory in Canadian agriculture. The trajectory not only suggests that, in practice, producers have adapted to multiple risks and opportunities in some ways more than others but also hints at the presence of a predominant interactive effect of multiple risks and opportunities in Canadian agriculture, at least in the recent past. This twofold argument is developed in the next section with the aim of informing our conceptualization of climate change impacts and adaptations in Canadian agriculture. Finally, a summary of the chapter's main arguments is presented.

At this point, it may be helpful to note that the discussion in this chapter is consistent with what Tol and colleagues (1998) label the "temporal analogue" approach to climate change impact assessment, which is that strand of re-search that sees utility in documenting *actual* adaptations to current cli-matic and non-climatic conditions (see, for example, Bradshaw et al. 2004; Chiotti et al. 1997; Smit et al. 1996; Smithers and Smit 1997b). As noted in Chapter 4, the focus on current conditions is also fundamental to vulner-ability assessments.

Conceptualizing Farm-Level Exposure and Adaptation to Multiple Risks

Smit and colleagues (1996) and Smithers and Smit (1997b) make use of an agricultural systems framework to conceptualize how producers experience climatic variability and change relative to other external stimuli. The frame-work, as conceived by Olmstead (1970, 32), starts with the recognition that "no farm exists unto itself." Rather, individual farms exist within, and hence are exposed to, a spatial hierarchy of biophysical, political, social, and eco-nomic influences, including, for example, inter-periodic climatic variability, interest rate shifts, changes in consumer preferences, or new environmental regulations (see Figure 8.1). The responses of many individual producers to these external influences can be highly variable given differing perceptions and sensitivities, both of which are likely a function of the particular attributes of individual farms and farm operators. Finally, according to the framework, farm-level decisions produce impacts for the individual, as well as aggregated economic, environmental, and social impacts, all of which can be expected to create system feedback.

The agricultural systems framework effectively *places* climate risks rela-tive to other risks of production, marketing, and finance in primary agricul-ture, and does so in a way that at least enables one to envision certain *interactions* among various external stimuli. That is, it makes it possible to see a producer's larger context within which any perceived change in cli-mate will be experienced and potentially addressed. The potential *means* by which a producer might address climate change, whether by changing out-puts or taking off-farm work, are also identified in the framework, although these are not itemized in Figure 8.1. Hence, the framework is comprehensive in the sense that it identifies: (1) possible determinants, climatic and other-wise, of farm-level decisions, including some potentially interactive ones; and (2) possible decisions that producers may make. Understandably, then, it is not particularly insightful in terms of identifying common decisions and their common determinants; its fundamental insight is that farm-level decisions to a variety of ends can happen for a variety of reasons. Place- and time-specific empirical applications of the framework, as carried out by Smit and colleagues (1996) and Smithers and Smit (1997a), can be highly enlighten-ing, offering insights into key external influences, and perhaps even multiple interacting influences, as well as typical producer responses. These findings

Figure 8.1

A generic farming systems framework

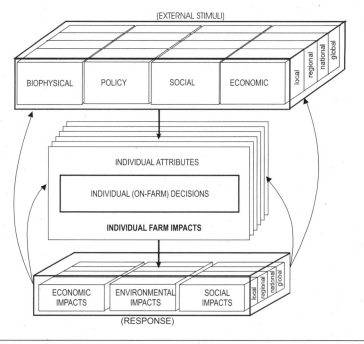

Source: After Bradshaw and Smit (1997).

are not fed back into the framework to enable its refinement into a model, however; subsequent empirical investigations begin with the same comprehensive but in some ways unilluminating framework.

The vulnerability approach to climate change impact assessment (see Chapter 4 and Part 4) has a similar quality. Its insights derive not from its theoretical premise, which posits that a producer's vulnerability is a function of his or her risk exposure and adaptive capacity, but rather from its application in particular places at particular times (see, for example, Chapter 12). Such applications, which make use of "bottom-up" data-gathering techniques, identify: (1) just those risks that are deemed problematic by producers; and (2) the specific means that producers possess to try to manage those risks. While this narrowing and categorization of the agricultural systems framework's external stimuli and individual attributes into factors of "exposure" and "adaptive capacity" represents a conceptual advance, the vulnerability approach shares the same fundamental characteristic of its predecessor; it is not a model that necessarily becomes more informative with each empirical application.

How might the above conceptualizations be bolstered and made more informative? One way to do this is by incorporating certain farm-level trends in Canadian agriculture, which suggest that, over the recent past, producers have regarded some adaptations to multiple risks and opportunities as more sensible than others, and perhaps that producers have deemed some signals to be more significant than others. These trends are identified in the next section, while their significance is explained in the section after that.

Canadian Agriculture's Historical Trajectory

Canada's primary agricultural sector is dynamic, as are most individual operations that comprise it. Further, the farm-level actions that create this dynamism are not random and directionless but rather purposeful and unidirectional. And lastly, to state the obvious, these actions represent adaptations by many thousands of producers to a variety of external risks and opportunities, including, but not limited to, climatic ones. The fact that primary agriculture is in a state of constant change is certainly not a new idea in the climate change impacts and adaptations literature, although it is one that is often downplayed in many impact assessments. A notable exception is Easterling's work (1996, 3), in which the author recognizes the need to identify significant trends in order to "project a plausible scenario of the future of North American agricultural production ... to establish a baseline against which to measure the consequences of climate change." Many of the trends recognized by Easterling (1996) are also identified here and elaborated on in a prairie context in Chapter 9. These include:

- increased output productivity among individual operations
- larger and fewer farms
- more specialized production on individual farms
- more intensive production on individual farms
- greater integration of farms into the agri-food system
- more "pluriactivity" among individual producers and their families.

Taken together, these trends arguably constitute a historical trajectory. They do not represent all agricultural operations in Canada, especially those comprising the growing but still marginal organic sector. Even within the conventional sector, there are certain regions in Canada that have bucked dominant trends, such as the Lower Fraser Valley of British Columbia, where producers have recently expanded, rather than contracted, the total number of crops in their operations (Statistics Canada 2005a). These examples, however, represent anomalies relative to the dominant trends below.

To many observers (e.g., Cochrane 1958; Marsden et al. 1986; Ward 1993), the tendency of farmers to increase their production levels year after year, and hence to greatly augment aggregate supplies of agricultural outputs, represents the most fundamental characteristic of primary agriculture. Ward (1993) goes so far as to suggest that, for farmers, forever increasing one's productivity has become assumed logic. It is no wonder, then, that overall wheat production in Canada increased from under 200 million bushels at the turn of the last century to over 1 billion by the end (Statistics Canada 2005b; Canadian Wheat Board 2003). Even at the global scale, total cereal production increased from 877 trillion tonnes in 1961 to 2,101 trillion tonnes by 2001, thereby elevating food production per capita by 30 percent over that time period (FAO 2003). Some of the means by which farmers achieved these productivity gains are reflected in three further dominant trends of primary agriculture.

Farms in Canada, as elsewhere, have become fewer in number and larger in size over the past half-century. In terms of farm numbers, the peak year in Canada was 1941, when 732,832 were identified in the agricultural census. Since that time, continual amalgamation of farm properties has reduced the number to just 246,923 as of 2001. Figure 8.2 demonstrates this trend for the period 1981-2001. In it, one can see a reduction in the slope of the trend lines between 1991 and 1996, which was read by some observers as evidence of an approaching equilibrium that would end this historical trend; naturally, the far steeper slope of the trend lines for the subsequent period 1996-2001 put such speculations to rest. Farm concentration is a trend in Canadian agriculture that is likely to persist for some time to come.

Increased farm-level output specialization has been another characteristic of Canadian agriculture, and agriculture in the western world more generally, for at least a half-century but especially since the early 1970s, when

Figure 8.2

Farm concentration in Canada, 1981-2001

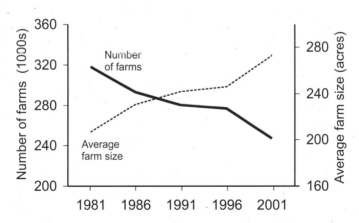

Source: Statistics Canada (2005b).

the Organization of Petroleum Exporting Countries (OPEC) oil crisis led to elevated grain prices, which prompted many farmers with mixed operations to drop livestock production in favour of producing just cereals (see Bowler 1985; US National Research Council 1989; Bollman et al. 1995; Ilbery and Bowler 1998). While this trend is widely acknowledged among agricultural analysts, farm-level evidence is often difficult to find (see Gregson 1996). For the Canadian Prairies, for example, it can be crudely identified by summing the number of farms reported in the census to be growing each of the principal crop types and dividing by the total number of crop farms. By this measure, the number of crop types grown per Prairie farm declined from 3.84 in 1971 to 3.00 by 1976 and just 2.91 by 1981. This trend is given further attention in the next section.

In addition to getting bigger and more specialized, productivity gains have been achieved through farm-level intensification, which is most clearly manifested in increased agri-chemical use. For example, fertilizer use in Canadian agriculture grew from just over 400,000 metric tonnes in 1961 to over 2,600,000 metric tonnes by 2002 (FAO 2003). While the rate of growth has certainly slowed recently compared with the 1960s and 1970s, reflecting more precise use of agri-chemicals, farmers continue to look to inputs like fertilizers to produce more output.

The increased integration of farms into a larger agri-food system over the past century has resulted from both changed practices among farmers, such as purchasing seeds rather than saving them, and the remarkable expansion of the agri-food system itself, most clearly seen in the expansion of the processed food and food services sectors. As a result, primary agriculture's share of the total GDP of the agri-food sector was no more than 14 percent as of 2004 (AAFC 2006). This trend is also manifest in the ever-declining share of retail food dollars captured by primary producers. To take an admittedly extreme example, in 1981, Ontario corn producers received just 5 percent of the total cost of a box of corn flakes breakfast cereal; by 2001, their share was down to just 3 percent (Martz 2004).

Notwithstanding the fact that farms have become larger, more specialized, and more intensive, but perhaps reflecting primary agriculture's declining share of retail food dollars, farm operators have had to become increasingly "pluriactive," which is the final trend identified here. (Following MacKinnon and colleagues [1991, 59], the term "pluriactivity" is typically used to describe "the phenomenon of farming in conjunction with other gainful activity whether on or off farm.") While activities such as agri-food tourism receive a great deal of attention in both academic and popular circles, the most common and least glamorous pluriactivity is off-farm work. Figure 8.3 identifies not only the significance of off-farm income within Canadian producers' total income for the period 1993-2000 but also its growth relative to stagnant farm incomes. Even larger farms,

Figure 8.3

Average farm and off-farm income among Canadian producers, 1993-2000

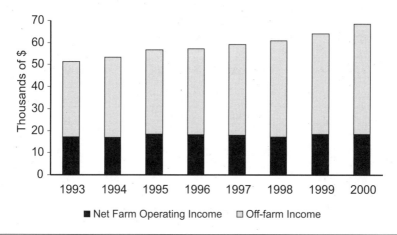

Source: AAFC (2000).

those with sales of $100,000 and more, are relying on off-farm income sources for about one-half of their family income (Statistics Canada 2000). For one of Canada's biggest agricultural sectors, beef, the figures for 2002, the latest year for which data are available, are particularly telling. The average total farm income among the 83,750 producers was $38,416, of which $26,437, or 69 percent, came from off the farm (Statistics Canada 2004).

Again, these trends capture the near-continuous practice of adaptation by many thousands of Canadian producers to a variety of external risks and opportunities, including, but not limited to, climatic ones. Their existence has important implications for the conceptualization of climate change impacts and adaptations within Canadian agriculture in at least two ways. These are identified in the next section.

Implications of the Trajectory for the Conceptualization of Climate Change Impacts and Adaptation in Canadian Agriculture

While the trends described above could be practically used to project a plausible future scenario of Canadian agriculture against which to measure the consequences of climate change, as suggested by Easterling (1996), these trends are of use for at least two conceptual reasons as well. Recall that the trends reflect producers' tendency over the past half-century to respond to multiple risks and opportunities, including climatic ones, in some ways more than others. This fact offers insights for that strand of research that seeks to conceptualize potential farm-level adaptations to climatic variability and change as noted in Chapter 2. In short, the suggestion is that potential

Figure 8.4

A conceptual model of the impact of Canadian agriculture's historical trajectory for farm-level decision making in light of multiple risks, illustrating the point at which a would-be adaptation to multiple risks is considered

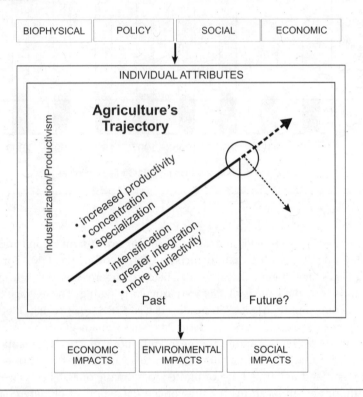

adaptations that are consistent with farmers' past practices are more likely to be implemented than those that constitute truly novel behaviour; the latter is certainly possible given extraordinary climatic conditions, but it is less likely. This is the nature of a *trend*. Figure 8.4 attempts to depict this phenomenon. Drawing upon the basic elements of the agricultural systems model, the diagram suggests that numerous exogenous stimuli, as perceived by individuals and influenced by their particular attributes, have produced agriculture's historical trajectory, manifested in such trends as farm concentration and intensification. Into the future, climatic stimuli will be experienced by producers alongside other biophysical and non-biophysical stimuli, leading to management decisions that *could* constitute a fundamental reversal in behaviour (i.e., the thin dotted trajectory), from, for example, farm enlargement to contraction; it is more likely, however, that these decisions will serve to continue historical trends (i.e., the thick dotted trajectory).

The findings and arguments of Bradshaw and colleagues (2004) support this point. In the context of mid-1990s policy reforms in Canadian Prairie agriculture, some analysts anticipated that producers would break from the long-time trend of crop specialization (i.e., they would expand the range of crops grown on their farms) in order to buffer increased market and climate risks. Bradshaw and colleagues (2004) sought to determine whether such a dramatic shift was indeed evident. The analysis was further motivated by a number of post-policy implementation studies that identified increased crop diversity in the Prairies, at least at the regional scale (see Campbell et al. 2002; Zentner et al. 2002). Based on a representative sample of individual Prairie producers for the period 1994-2002, it was determined that farmers had not diversified their cropping patterns; rather, the trend towards specialization, as measured by the number of crop types per farm, continued at an average rate of –1 percent per year. The authors did not exclude the possibility of a break from the specialization trend in the future given certain changed conditions, including but not limited to an increased observance of crop-damaging extreme weather events, but cautioned that this was unlikely given the persistence of other external stimuli. These include, for example, the marginal cost advantage of increasing scales of production; the relative affordability and effectiveness of agri-chemicals and single-function machinery; pressure from agribusiness, creditors, and, certainly during the 1970s, government extension officers; and the provision of below-market-cost crop insurance (Bollman et al. 1995; Gertler 1999; Gregson 1996). In short, Bradshaw and colleagues (2004) effectively revealed the illogic of a climate change adaptation strategy commonly identified in agriculture – crop diversification – by placing it within the context of agriculture's historical trajectory and one of that trajectory's key dominant trends in particular.

Recognition of persistent trends and the factors that appear to drive them, as exemplified by the case of specialization, also compels us to consider the possibility of predominant interactive effects of multiple risks and opportunities in Canadian agriculture. That is, rather than conceptualize primary agriculture as characterized by highly variable farm-level decisions that are a product of any number of random combinations of external stimuli and internal attributes (suggested by the agricultural systems framework in Figure 8.1), it may be more reasonable to conceptualize primary agriculture as characterized by particular farm-level decisions that are the product of a limited number of particular combinations of specific external stimuli. Consistent with Troughton (1991) and Le Heron (1993), it might even be reasonable to conceive of primary agriculture as characterized by highly homogeneous farm-level decisions, which are the product of one predominant external signal – depreciating agricultural commodity prices.

Given highly inelastic demand for foodstuffs, the slow growth of total demand, and rapid technological change that increases agricultural commodity supplies over time, agricultural commodity prices tend to be highly variable and, in inflation-adjusted terms, decline over time (Rausser and Hochman 1979). This tendency is reflected in the price data for Canadian wheat over the last century (Figure 8.5). Wheat prices in 1991, the low point for the century, were just 14 percent of their 1919 values, the high point for the century. Further, recent prices appear to be far from anomalous; a long-term trend line is discernible. This deflationary tendency might not be deemed troublesome by farmers were their input costs also to deflate, but this is not the case. Because the cost of farm inputs tends to increase faster than the price of farm outputs, farmers regularly find themselves squeezed into a position of declining profits. Further, there is considerable evidence (e.g., Canadian Royal Commission on Farm Machinery 1969; Friedmann 1995; Labao 1990; Rochester 1940) that this "squeeze" is exaggerated by the oligopolistic behaviour of agribusiness, both upstream and downstream of the farm (see Chapter 9). This is one obvious implication of farmers' increased integration within the agri-food system. The results of the 2001 Canadian agricultural census are telling in this regard. In 1995, total operating expenses equalled 83 percent of total revenues; by 2000, they had climbed to 87 percent. At the extreme end, cattle farms spent 94 cents for every dollar in sales in 2001 (AAFC 2006).

Figure 8.5

The real price of wheat in Canada, 1914-98

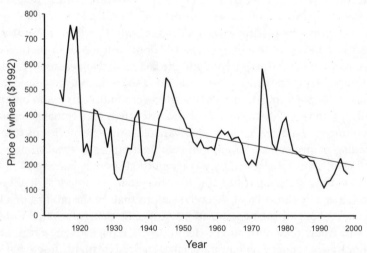

Source: Canadian Wheat Board (2003).

Farmers' predominant response to this profit squeeze has been to attempt to increase their productivity via farm enlargement, intensification, and output specialization, and in this they have been remarkably successful. Their success contributes to larger surpluses, which in turn contribute to further depressed prices and the need for further productivity gains. While collectively this may seem illogical, for most individuals there is no other logic. Again, the 2001 Canadian agricultural census is revealing. It found that the largest farms in Canada had the best ratio of expenses to sales, whereas the smallest farms had the worst. Farmers know that they must expand their operations to maintain their incomes. Even so, incomes have not been sufficiently maintained; hence, we see increasing reliance on off-farm income sources among farmers.

In short, while explanations of primary agriculture's trajectory can draw upon a variety of exogenous factors, depreciating commodity prices alone can reasonably account for many if not all of the trends identified earlier. Of course, commodity prices will continue to play a formative role well into a climatically different future. Indeed, the (Canadian) National Farmers Union (2003, 8) goes so far as to suggest that the ultimate impact of climate change upon agriculture will depend on commodity prices. In a brief to a Senate committee investigating the potential implications of climate change for Canadian agriculture, it argued: "To adapt to any changes, farmers need prices high enough to give a farm family long-term stability and give it capital to invest. Fair and adequate commodity prices are essential if our farms are to adapt to climate change."

Summary

Canadian farmers have long contended with, and adapted to, a range of risks and opportunities, be they extreme weather conditions, shifts in government policies, or variable and deflationary commodity prices. The transformation of primary agriculture in Canada over the last five decades – evident in the trends towards greater productivity, concentration, speciali-zation, intensification, integration, and pluriactivity – attests to this. This chapter has sought to incorporate this fact into our conceptualization of climate change impacts and adaptation in Canadian agriculture. Prior con-ceptual efforts (e.g., Smit et al. 1996; Smithers and Smit 1997a) have use-fully drawn upon Olmstead's agricultural systems framework (1970) to reveal the place of climate risks relative to other risks, the possible interactions among multiple risks, and even the various means by which a producer might address climate risks. While broadly informative, the framework does not point to likely drivers of farm-level decisions and likely outcomes. Hence, we have argued that our conceptualizations need to incorporate, or at least acknowledge, primary agriculture's historical trajectory. What would this imply for our understanding of climate change impacts and adaptations in Canadian agriculture?

From the most to the least intuitive, three insights are offered here. First, *it can help us to see how climatic stimuli, in tandem with many others, have been experienced and responded to by farmers in the recent past.* The many observable trends of primary agriculture reflect producers' tendencies to adapt to multiple risks and opportunities, including climatic ones, in some ways more than others (e.g., through specialization). More so, the remarkable homogeneity of farm-level behaviour suggests a predominant interactive effect of multiple risks and opportunities in Canadian agriculture; here, the significance of commodity prices and the historical cost/price squeeze is obvious.

Second, *it should compel us to rephrase the question of how climate change might impact upon agriculture.* Perhaps the more relevant question is: How might climate change impact upon primary agriculture's current trajectory? Or, put another way, by adhering to this trajectory – that is, by continuing to expand, specialize, intensify, integrate, and seek wider income sources – will farmers continue to manage multiple risks and seize multiple opportunities, or will producers' necessary response to climate change represent a break from this trajectory? A conservative gambler would probably pick the former.

This is the third insight offered by the recognition and conceptual incorporation of primary agriculture's trajectory. *It suggests that future responses by producers to climatic risks and opportunities will probably mirror past responses, as potential climate change adaptations that fail to conform to long-established farm management plans (and even longer established farming cultures) will likely be viewed as illogical.* This was certainly true of crop diversification, a widely recognized potential climate change adaptation strategy, in the Canadian Prairies over the 1990s (Bradshaw et al. 2004). While it is not impossible for future practice to break from past practice given particular climatic conditions, it is improbable. To return to the simple fact from which this chapter set out, climate represents just one of many sources of risk (and opportunity) to which farmers are exposed and respond.

Notes

1 Obvious recent examples in Canada include the loss of the US export market for live cattle due to the discovery of BSE (bovine spongiform encephalopathy) on a few Alberta farms, and the associated decimation of prices for feed crops like corn and soybeans.

2 A common approach in much of this climate-focused research has been to control for other variables through declarations of *ceteris paribus*. Recognition of the need to break from what Wenger and colleagues (2000) label the "single-stressor, single-endpoint" paradigm is growing, however.

9

Biophysical and Socio-Economic Stressors for Agriculture in the Canadian Prairies

Henry David Venema

This chapter considers the historical and projected climatic and socio-economic stresses in the Canadian Prairies with respect to the concept of resilience. (As noted in Chapters 4 and 14, the term "resilience" is used synonymously with "adaptive capacity" since both refer to a system's ability to respond to stress and impacts.) An emerging body of international development literature indicates that agricultural programs and policies have typically not accounted for the implications of high climatic variability. Climatic variability, and particularly drought, have long characterized the prairie agri-ecosystem (see Chapter 6). Paleoclimatological records indicate episodes of more extreme drought than those found in the instrumental record. Future climate change may increase the risk of aridity and drought.

The cumulative impact of drought in the context of prevailing agricultural practice in the Canadian Prairies has resulted in repeated crop failures throughout the twentieth century, many of which have required costly government bailouts. Current agricultural policies promote intensified production for export markets but have not succeeded in improving farm incomes, which remain stagnant and a great source of strain in rural communities. A major policy change has been the cessation of the Crow Rate subsidy[1] on grain transportation, which was intended to promote more diverse Prairie agriculture. The end of the Crow Rate led to the rise of intensified livestock operations (ILOs) that could utilize the initially low-cost grain. ILOs have in turn aroused public ire because of their perceived negative health and environmental impact, particularly with respect to water quality and quantity – concerns that are ever more acute in the context of climate change-induced future water scarcity. The recent bovine spongiform encephalopathy (BSE) crisis that has devastated the livelihood of cattle producers highlights some of the major vulnerabilities of the current export-oriented production system.

Simultaneously, the introduction of genetically modified (GM) canola has been a commercial success but destroyed export opportunities for organic canola producers because of seed supply cross-contamination. Organic farming was anticipated as a diversification opportunity for Prairie agriculture consistent with sustainable development, but growth has been very modest. The Canadian Wheat Board and organic producers are currently resisting the introduction of GM-wheat, primarily because of fears that GM-contaminated wheat shipments would compromise Canadian wheat export markets and limit organic farming opportunities.

Prairie agriculture does not currently possess the attributes of resilience; it cannot absorb shocks and stressors without large external aid and capital infusions to restabilize it, and the increased risk of future climatic stresses and shocks likely exacerbates this situation. The advent of climate change may, however, provide a renewed rationale for rethinking Prairie agriculture from the perspective of socio-ecological resilience.

This chapter reviews the socio-ecological and economic conditions now existing in the Canadian Prairies and examines the role of recent policy decisions in creating those conditions. It concludes with some observations on how climate change may force return to resilience-based agricultural policy such as that witnessed in the mid-twentieth century with the advent of the Prairie Farm Rehabilitation Administration (PFRA) and the Canadian Wheat Board.

Socio-Ecological and Economic Conditions in the Canadian Prairies
In November 2003, the Standing Senate Committee on Agriculture and Forestry concluded that Canadian agriculture would be affected by climate change, noting that more frequent and widespread drought on the Prairies was expected (SSCAF 2003). The Prairies produce well over half of the total value of Canadian agri-food exports (Tyrchniewicz and Chiotti 1997) but are still frequently affected by climate-related disasters. The prospect of more frequent and larger droughts afflicting the Prairies only compounds the multitude of shocks and stressors with which Prairie agriculture has had to cope, including stagnant commodity prices, closure of the American border to beef exports, and further threats to export markets from the introduction of genetically modified wheat.

Chapter 6 summarizes the impacts that climate change is having and might have on Prairie agriculture and the rural communities involved. It notes that the history of Prairie agriculture is characterized by adaptations to climatic variability, including the development of drought- and frost-tolerant crops and farming practices that conserve soil moisture. Considerable evidence exists, however, that more aggressive policies to increase coping capacity may be required. The drought of the Great Depression provides a classic example of maladaptation; the 1930s drought affected 7.3 million acres and forced the distress migration of a quarter of a million people (Goodwin 1986). This mass exodus did bring about a major institutional response, namely, the formation of the Prairie Farm Rehabilitation Administration, a federal government agency mandated to promote soil and water conservation. The impacts of subsequent droughts have largely been simply absorbed by a much larger national economy that is less dependent on agriculture.[2] The 1984-85 drought, for instance, affected most of the southern Prairies, and cost Canada at least a billion dollars of GDP (Ripley 1988).

In the severe drought year of 1988, agricultural export losses topped $4 billion. Despite assistance payments of over $1.3 billion, Manitoba showed net farm income losses of 50 percent and Saskatchewan 78 percent, and an estimated 10 percent of farmers and farm workers left the agricultural sector that year alone (Herrington et al. 1997). Even in 1991, a year of record high wheat production, emergency payments (i.e., above regular assistance and insurance programs) were still in excess of $700 million (Sauchyn and Beaudoin 1998). Recent analysis of the 2001-02 drought indicates losses of $3.6 billion and $5.8 billion in agricultural productivity and gross domestic product, respectively, which was manifested as the loss of 41,000 jobs in the agricultural sector (Wheaton et al. 2005).

The cumulative toll on Prairie land resources from stresses related to climate and conventional agricultural practice is also telling. In 2000, the PFRA released a report entitled *Prairie Agricultural Landscapes: A Land Resource Review* (PFRA 2000), which concluded that: (1) more than 50 percent of fields with annual crops are threatened by erosion and require intensified use of crop residues and permanent cover to maintain soil health, and (2) 14-40 percent of soil organic matter (the vital component of the soil fabric, responsible for improving soil structure, tilth, fertility, and health) has been lost from Prairie soils since cultivation began. Given this situation, serious questions are being raised about how the family farm on the Prairies is coping with these agri-ecological stresses and, equally importantly, with severe concurrent economic stress. The short answer is, with grim determination to hang on.

Ninety-eight percent of farms in Canada are still family-owned and operated, but in 2000, 73 percent of the income of the average farm family came from off the farm. Low commodity prices are forcing farm families into the stressful existence of combining full-time jobs with farm management. Many simply cannot afford to go on, and between 1996 and 2001 the number of farms in Canada declined by 10.7 percent (Martz 2004). A story in the *United Church Observer* (Driver 2004, 27-29) captures the outcome of these cumulative stresses on the social fabric of Prairie communities:

> Continuing drought, low grain prices, mad-cow disease, grasshopper infestations, and flu-infected poultry: these are nail-biting times in rural Canada. Some farmers are working two or three jobs to make ends meet. The extra work means they can't volunteer for church positions and activities.

And they have less money to put on the offering plate. "People are cutting back everywhere, including the church," says Rev. John Lea of the Assiniboia pastoral charge in southern Saskatchewan ... Auctions of entire farms are routine now, but the age of the owners is creeping downward to include 40-year-olds, says Lea. The stresses are bursting the entire community and many more families are moving away from small towns to cities. In the last seven years more than 500 people, or about one-sixth of the population have left Assiniboia ... for farmers like Donna Zimmer, there really is no other choice but to keep on going. "What else can you do?" she says. "I have worked off the farm all of our married life. I'm looking again for work because we are not making it. We're just hanging on, waiting for something to turn around.

Just how bad is the farm income crisis? Figure 9.1 compares total farm receipts with farm debt and reveals the widening debt burden faced by Canadian farmers. The stagnation of commodity prices relative to processed food prices explains much of the farm income crisis. According to the National Farmers Union (NFU) (2005), producers have continued to receive a smaller share of the total food dollar since the mid-1970s, with higher returns going to retailers and food processors.

The NFU argues that retailers have been using their market power to simultaneously inflate consumer prices and push down prices to farmers (and

Figure 9.1

Total farm debt and farm receipts, 1981-2001

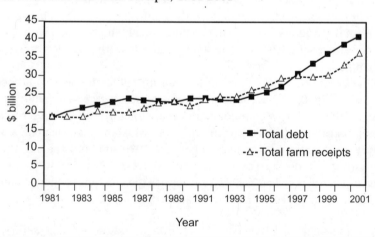

Source: National Farmers Union (2005).

wages to workers). In contrast, both Martz (2004) and the NFU (2005) note the relative stability of the supply-managed dairy, poultry, and egg sectors, where marketing boards match supply with demand.

The NFU goes on to suggest that Canada's export-driven agricultural trade policy[3] has failed the family farm, asserting that if one were to list the agricultural sectors most heavily focused on export (grains, oilseeds, and hogs) and the sectors hardest hit by the farm income crisis, you would have the same list. In contrast, the sectors that focus on supplying the Canadian market – dairy, eggs, and poultry – have largely escaped the crisis. The NFU indicates that while Canadian agri-food exports have expanded fivefold since 1979, family farm incomes have declined over the same period (see Figure 9.2).

The federal role in setting agricultural policy and thus influencing socio-economic and environmental outcomes is unequivocally huge. The following section surveys the fundamental forces and actors currently shaping agriculture policy and practice in the Canadian Prairies. We conclude that federal agricultural policy requires much more vigorous focus on building community-scale resilience to mitigate the compounding effects of agricultural trade liberalization and climate change.

Figure 9.2

Agricultural exports versus farm income, 1970-2003

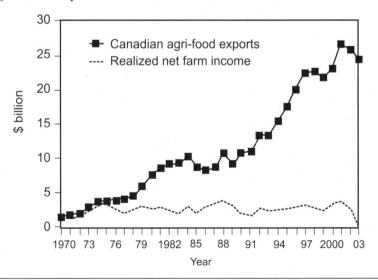

Source: National Farmers Union (2005).

Prairie Agricultural Policy Issues

The federal government's restructuring of agriculture in Canada since the 1980s is similar to that accomplished by International Monetary Fund (IMF)/ World Bank in the structural adjustment of economies in the developing world. The key instruments in domestic structural adjustment have been the World Trade Organization (WTO) and the North America Free Trade Agreement (NAFTA). Policies reminiscent of structural adjustment that were adopted in Canada and applied to the agricultural sector include: a focus on production for export; dramatic cuts in government spending; deregulation; measures to attract foreign investment; privatization of government industries and utilities; removal of farm subsidies, price controls, and other supports; and implementation of a freely floating currency.

By 1989, according to the NFU, Canada no longer had an agricultural policy as such, but instead had "a trade policy that masquerades as farm policy" – a very successful trade policy nonetheless. Agri-food exports doubled in seven years from $10 billion in 1989 to over $20 billion in 1996. Inducements to foreign investment have worked rather well also; by 1999, one US transnational, Archer Daniels Midland, owned almost 50 percent of Canadian flour milling capacity.

Between 1991/92 and 1999/2000, government spending on agriculture dropped by 52 percent, from the peak of over $6.1 billion in 1991/92 to approximately $2.9 billion for 1999/2000. According to the NFU, however, no policy decision has had a greater negative impact on western farmers' income than the ending of the Crow Rate benefit, which subsidized shipping costs. The agriculture minister at the time, Ralph Goodale, claimed that ending the Crow Rate would: (1) help diversify Prairie agriculture, (2) boost the value-added sectors, and (3) bring Canada into compliance with international trade obligations to reduce subsidies.

Moving grain to port for export is now a cost borne entirely by the farmer, and on average is about 25 percent of the total cost of the farm operation. The NFU claims that eliminating the Crow Rate has devastated grain farmers' gross incomes – reducing them by as much as 40 percent through increased rail costs. Simultaneously, the rural rail network has been greatly reduced, and with it the number of grain delivery points. Since the beginning of the 1999-2000 crop year, the number of licensed primary and process elevators located in western Canada has fallen from 1,004 to 416, a reduction of 59 percent. Although the railways have captured significant efficiency gains (and gained record profits), farmers have been forced to pay increased costs for trucking, and rural communities have shouldered rising tax burdens associated with increased road maintenance.

Some of the diversification promised by Minister Goodale with the elimination of the Crow Rate did take place – not through a return to relatively

low input classical mixed farming but rather through new forms of specialized production as farmers adjusted to new economic realities. Pork and
beef production are prominent examples of intensified specialized production. Between 1991 and 2001, Canadian beef exports increased almost fivefold, much of it concentrated in southern Alberta. This unfettered growth
came to an abrupt end in May 2003, when the United States closed its border to Canadian beef after the discovery of a single case of bovine spongiform
encephalopathy (BSE), or mad cow disease, in northern Alberta. Even if the
BSE crisis had not dramatically curtailed beef production and export, climate change and water resource limitations might have placed hard constraints on the industry (de Loë 2005). Sixty percent of Alberta's beef
production takes place in the irrigated areas of the province, where moisture deficits are highest and are projected to increase with climate change
(de Loë 2005). The economic impacts of the BSE crisis have been exacerbated by the consolidation and closure of slaughtering capacity in Canada
throughout the 1980s and 1990s, a trend concurrent with the liberalization
forces that reshaped the meatpacking industry.

Trends and drivers in the pork industry are broadly similar. For example,
between 1990 and 2000, the number of hog farms in Manitoba has declined more than 50 percent, from 3,150 to 1,450, while the average number
of hogs per farm has more than tripled, from 388 head to 1,290 head. Overall hog production has increased rapidly, topping 4.8 million in 1999, 89
percent of which is exported. The rapid growth in intensive hog operations
is attributed to:

- changes in world grain trade resulting in relatively static volumes of grain
  being sold at ever declining prices (constant dollars) due to technology
  improvements
- loss of the Crow Rate benefit on export grain, resulting in farmers facing
  the full freight bill and lower (at least initially) feed grain prices
- growth in world demand for meat due to rising incomes.

This heavy intensification in the hog sector, and its concentration in certain locations within the province, has heightened public concerns regarding the environment, particularly air and water quality and public health.
A broad-based coalition of environmental non-governmental organizations
(NGOs) charge that hog ILOs threaten both water supplies and water quality,[4] and are particularly opposed to the practice of field application of liq-
uid hog manure, which is very high in nitrates and phosphorus. Noted
University of Alberta ecologist David Schindler warns that the combination
of declining streamflows due to climate change, and the concentration of
nitrates, phosphates, and pathogens from ILOs (some of which produce as

much waste as medium-sized cities), will seriously imperil the freshwater resources on the Prairies (Schindler 2001).

In the late 1990s, observers of Prairie agriculture anticipated that organic farming rather than industrial beef and pork production would be a major component of diversification anticipated by Minister Goodale at the time the Crow Rate was eliminated (Gertler 1999). The Organic Trade Association defines organic farming as "an ecological production management system that promotes and enhances biodiversity, biological cycles and soil biological activity. It is based on minimal use of off-farm inputs and on management practices that restore, maintain and enhance ecological harmony."

On the basis of a twenty-two-year comparison of organic and conventional grain-based farming systems, Pimental and colleagues (2005) conclude that the benefits of organic agricultural production are higher soil organic matter and nitrogen, lower fossil energy inputs, yields similar to those of conventional systems, and – critical to climate change adaptation objectives – conservation of soil moisture and water resources (especially advantageous under drought conditions).

Although interest among Canadian farmers in achieving certified organic status was initially strong, Statistics Canada indicates that organic farming is increasing only slowly. Saskatchewan, the province most impacted by the loss of the Crow Rate, leads the way, with 773 of the 2,230 certified organic farms in Canada[5] (about 1 percent of all farms in Canada). Between 2000 and 2003, the number of fruit and vegetable farmers claiming to be using organic production methods increased from 640 to 660.[6]

Whether by design or accident, agribusiness forces have played a role in limiting the growth of organic agriculture. The case of Roundup-Ready Canola is a particularly interesting one. Roundup (glyphosate) is a proprietary herbicide marketed by Monsanto. Roundup-Ready (RR) Canola is a GM variant that has specific resistance to glyphosate. The pairing of Roundup and RR Canola is marketed as a technology package to promote higher yields and soil conservation since the herbicide minimizes conventional tillage requirements to control weeds (Gertler 1999). Conservation tillage (whether or not associated with GM technology) conserves soil and water resources and is part of the portfolio of agricultural practices generally regarded as good practice for adapting to climate change, since it offers protection from both flood and drought impacts (Wall et al. 2004).

Monsanto began selling its RR Canola seed in 1996. By 1998, buyers were asking farmers for GMO (genetically modified organism)-free certificates to protect the integrity of their product for premium export markets that demanded a GMO-free canola. Within several years, however, it became apparent that GMO-free canola seed was no longer available because cross-contamination with non-GM canola had become too extensive.[7] The increasing level of contamination of organic crops led to certified organic farmers abandoning canola as a crop, as it became impossible to guarantee a product free of GMO contamination. Nonetheless, the market for Canadian canola remains strong; Canada is still the biggest single country producer of the crop. About 70 percent of Canadian canola is GMO. The Canola Council[8] claims that GMO varieties provide benefits to the grower, the industry, and the environment from a reduction in the required tillage operations compared with conventional variety growers, and from improved soil conservation and reduced herbicide use. Major importing countries include Japan and China, neither of which imposes any labelling restrictions on GMO canola. In contrast, the European Union does impose labelling restrictions on GMO canola, effectively eliminating this export market for organic producers who cannot acquire certified GMO-free canola seed. This fledgling but lucrative market for organic canola producers was destroyed.

Meeting Climate Change Challenges for Sustainable Agriculture in the Canadian Prairies

In a special issue of the Institute of Development Studies (IDS) *Bulletin on Climate Change and Development,* Scoones (2004) emphasizes the historical disconnect between the biophysical reality of fragile agri-ecosystems and the techniques and policies intended to manage them. Scoones focuses specifically on the pastoral rangelands of Africa, describing them as regions "where systems are not at equilibrium, where sometimes chaotic, often stochastic, dynamics prevail and where predictability and control are false hopes" (114). These types of regions, where equilibrium conditions do not apply, exist in very large swathes of Africa. Among their climatic conditions is a coefficient of variation of rainfall that is more than 30 percent – a degree of variability similar to that experienced in the Canadian Prairies (and projected to increase).

Scoones observes that the dominant management prescriptions in these African regions have emphasized developing a more orderly, predictable form of livestock production amenable to the market and external management. Such initiatives continually failed, however, and were found to be "wildly inappropriate" for the uncertain conditions typical of the area. Scoones recognizes that if climatic uncertainty and variability are on the rise globally due to climate change – creating non-equilibrium conditions such as is currently the situation in the dryland regions of Africa and quite conceivably in Prairie Canada – then "we must shed our blinkered equilibrium views and solutions and search for alternatives that allow for 'living with uncertainty'" (Scoones 2004, 116). As demonstrated in this chapter and in others (e.g., Chapters 8 and 10), this challenge resonates strongly in the Canadian Prairies. The economic vagaries related to agricultural trade liberalization and the shocks and stresses associated with climate change ensure that the only certainty is more uncertainty and unpredictability.

Reforming agricultural policy in the Canadian Prairies for sustainability in light of a changing climate can benefit, however, from insights from resilience theory, which explicitly assumes uncertainty and unpredictability. As explored more fully in Chapter 14, resilience theory began with the work of C.S. Holling (1973, 1986, 2001) and the observation that complex adaptive systems, including socio-ecological systems, exist as nested hierarchies of systems of various scales and types. Holling notes that such systems are subject to unpredictable and sometimes sudden changes, but through their capacity for self-organization, they can buffer and adapt to disturbances up to a point.

Resilience is the goal when managing complex adaptive systems. Resilient systems are said to be those that can tolerate external shocks and surprises and have the capacity to buffer perturbations, to renew and reorganize after change, and to learn and adapt in a dynamic world. Resilience theory leads to a conclusion similar to that drawn by Scoones (2004): that management techniques that use rigid control mechanisms to seek stability inevitably undermine resilience and enhance the breakdown of socio-ecological systems (Berkes et al. 2003).

The implication for Prairie agricultural policy is that there is no silver bullet for dealing with the compounding stresses of agricultural trade liberalization and climate change, but that "no-regrets" strategies such as the maintenance of livelihood options and ecological integrity should be paramount policy foci. An appropriate litmus test for agricultural policy related to building adaptive capacity for managing climate change risks thus becomes: *Does the policy keep existing livelihood options open and does it create new options?* In the case of GM crops, for example, until a way can be found to prevent contamination of non-GM seed, the extreme risk to other livelihood options (such as organic farming) needs to be weighed in the balance. Conversely, eliminating the Crow Rate benefit was too blunt a policy instrument to produce a significant shift towards organic farming, and instead produced a new (and wholly unexpected) set of ecological risks associated with greatly expanded industrial livestock operations. The rapid growth in ILOs then created high vulnerability to trade shocks such as those from BSE – a consequence of a hardwired continental trade system – and has increased the risks of water quality impairment, a condition expected to worsen with climate change.

One potential policy "prong" that bears much promise for improving the resilience of Prairie agriculture to climate change and other stresses is payment to farmers for providing ecosystem goods and services (EGS). Building up soil conditions, or improving the health of the soil, is a key aspect of ensuring agri-ecosystem resilience. Paying farmers for specific, well-defined soil conservation measures is a policy option that will enhance soil health and thereby resilience, while at the same time broadening the base of a producer's livelihood options. Improving soil conditions by increasing organic matter means creating better conditions for withstanding moisture stress (whether too much or too little) directly related to weather and climatic conditions.

Although the theoretical basis for valuing EGS is incomplete, an early market exists in the form of payment for carbon sequestration when fragile agricultural land prone to wind and water erosion is converted to permanent green cover crops.[9] Other producers' groups have proposed a more extensive accounting and payment system for EGS on primarily agricultural land.[10] EGS instruments could also be made consistent with the promotion of organic agriculture, in which case – given the reduction in greenhouse gas emissions and the soil and water conservation benefits of organic relative to conventional production documented by Pimental and colleagues (2005) – EGS instruments could promote climate change mitigation and adaptation simultaneously.

Prairie agriculture does not currently possess the attributes of resilience; it cannot absorb shocks and stresses without large external aid and capital infusions to restabilize it. The costs of continued crop losses, crop insurance payouts, and agricultural income stabilization programs should be carefully weighed against the cost of EGS payments that promote resilience. The need to adapt to climate change may be the crisis that forces this reckoning and demands institutional innovations comparable to the fundamental restructuring that occurred in the 1930s and 1940s with the advent of the Prairie Farm Rehabilitation Administration and the Canadian Wheat Board.

Notes

1   The Crow Rate was created in 1897 as a rate-control agreement (a regulated tariff) for the transportation of grain produced in the Canadian Prairies. It initially supported railway expansion in western Canada at the turn of the last century and helped reduce costs for grain producers. Over time it went through a series of alterations, eventually being called the WGTA (Western Grain Transportation Act). A number of factors led the Canadian federal government to end the WGTA in 1996 (Swanson and Venema 2006).

2   In 1999, approximately 2 percent of Canada's GDP was derived from agriculture: http://www.wd.gc.ca/rpts/audit/wdp/3_e.asp.

3   In 1993, federal and provincial governments set an ambitious target of doubling agri-food exports to $20 billion by 2000. Having accomplished their goal by 1996, well ahead of schedule, federal and provincial ministers pledged to redouble exports, to nearly $40 billion (4 percent of world agri-food exports) by 2005. The National Farmers Union claims that the latter goal was actually put forward by the Canadian Agri-Food Marketing Council, a private sector group that includes representatives of Maple Leaf Foods, Cargill, and McCain Foods.

4   See, for example, http://www.hogwatchmanitoba.org and http://www.beyondfactoryfarming.org.

5   http://www.cbc.ca/news/background/agriculture/subsidies.html.

6   http://www.statcan.ca/Daily/English/050428/d050428c.htm.

7   http://www.producer.com/articles/20021114/news/20021114news06.html. The Agriculture and Agri-Food Canada study, *Isolation Effectiveness in Canola Seed Production* by R.K. Downey and H. Beckie, disclosed that growers producing certified canola seed for the conventional

canola market in 2000-01 could not prevent genetic contamination of their seed by genetically modified (GM) canolas. The contamination was so severe that the authors recommended that four varieties sold in the conventional canola market be withdrawn or Breeder
and Foundation seed sources for the varieties be cleaned up: http://www.saskorganic.com/
oapf/pdf/canolastudy.pdf.

8   http://www.canola-council.org/overview.html.
9   There is a prototype program, "GreenCover Canada": http://www.agr.gc.ca/env/greencover-
    verdir/index_e.phtml.
10 For example, the Keystone Agricultural Producers (KAP) in Manitoba promote an EGS scheme
    referred to as Alternative Land Use Services: http://www.kap.mb.ca/alus.htm.

10

Institutional Capacity for Agriculture in the South Saskatchewan River Basin

Harry P. Diaz and David A. Gauthier

The adoption of adaptation strategies for climate change is particularly critical in regions that are potentially at greatest risk from or vulnerable to the impacts of climate change.[1] Dryland areas subject to water shortages, such as the South Saskatchewan River Basin (SSRB) in the Canadian Prairies, are an example of such vulnerable regions. As noted in Chapter 6, forecasted impacts of climate change on the regional water supply in the dryland areas of the Prairies suggest that adaptations are essential to optimize the benefits of water and to reduce economic, environmental, and social threats associated with scarce water resources.

This chapter focuses on the role of public institutions as a component of the adaptive capacity of rural communities and rural households for dealing with risks from changing climatic conditions and resource scarcities. Formal institutions are expected to serve the needs of civil society and thus it is important for those institutions to develop their capacity to implement activities in an environmentally sustainable manner and be held accountable through reporting on the sustainability of their activities.

The discussion here is intended to contribute to the understanding of the term "adaptive capacity" by discussing conceptual and methodological issues related to institutional adaptation to climate change. Research underway in the SSRB is used for illustrative purposes. This chapter is divided into two sections. The first sets up a framework for discussing and analyzing the adaptive capacity of institutions and presents those components of "institutional adaptive capacity" that are considered fundamental for supporting the adaptive capacity of agricultural producers and rural communities. Based on that analytical framework, the second section discusses the institutional scenario existing in the South Saskatchewan River Basin and its limitations and potential for the development of a regional adaptive capacity.

Institutional Adaptation to Climate Change

Already a region of water scarcity, the South Saskatchewan River Basin is expected to be seriously impacted by climate change as a result of reduced streamflows and water recharge and increasing evapotranspiration, with resultant severe impacts upon crops, livestock, and local ecosystems. The CCIAD (Climate Change Impacts and Adaptation Directorate) (2002) has summarized the potential impacts of climate change on water resources for the Prairies. These impacts include:

- changes in annual streamflow (possible large declines in summer), with implications for the different water users
- increased aridity and likelihood of severe drought, with losses in agricultural production and changes in land use
- increases or decreases in irrigation demand and water availability, with uncertain impacts on groundwater, streamflow, and water quality.

While some climate change models predict increased rainfall and snowfall in the Prairie provinces as a result of higher temperatures, a much greater loss of water by evaporation is also expected, resulting in overall drier conditions. That is, the major impacts of climate change on the Prairie provinces are expected to be loss of soil moisture and surface water. Furthermore, snow and ice are the principal sources of runoff that supply the Prairie lakes, rivers, and streams. Chapter 6 described a predicted decrease in snow accumulation in Canada's mountain ranges due to warmer winters, which will affect the availability of water for the Prairies.

In a river basin with a complex variety of physical and social conditions, the specific social impacts of climate change are likely to be heterogeneous. A likely common denominator, however, could be potential increases in water-use conflicts between sectors and within sectors, and between regions and users. With increasing aridity, the future demand for water will have to be considered against declining water availability, and competition for water between sectors will likely increase. Expected increases in industrial development and the expansion of urban centres such as Calgary will place additional pressures on rural area water supplies. In the agricultural sector, farmers and ranchers are already being pressured to increase production in a context where they face increasingly unpredictable supplies, and must increasingly compete with cities and other economic sectors for available water. Conflicts within and among sectors – as in the case of conflicts around irrigation – are likely to increase.

Within the context of predicted water scarcities, the need to understand regional climate change adaptive capacities is fundamental. There is no doubt that developments in technology and infrastructure, as well as the availability of economic resources, will be essential to improve water-use efficiency. These capacities, while necessary, are likely to be insufficient, however. As the Food and Agriculture Organization of the United Nations argues, "institutional changes are going to be as important as, or more important than, technological ones" (FAO 2003, 372). The institutional changes that will be required are those involving the development and implementation of comprehensive support mechanisms that improve the capacity of different sectors to adapt to climate change.

Of paramount importance is the development of adaptive mechanisms in those human settings that are the most vulnerable to climate variability and change, such as in rural communities and rural households. (Chapter 15 explores such adaptive capacity in relation to rural families and communities through his analysis of the Dust Bowl migration in the southwestern United States.) These adaptive mechanisms will involve capacities such as human and social capital, access to information, and availability of and access to resources. Chapter 3 has pointed out that institutional actions at the level of the state, able to support the strengthening of mechanisms in communities and households, are of prime importance if they are to adapt to climate change. Although not so overtly exposed to the most direct effects of climate change, those formal institutions are fundamental to the adaptive capacity of rural communities and agricultural producers. What they do or do not do impinges directly upon human communities and ecosystems, which are vulnerable to varying degrees of exposure to stress. Responsive adaptations will involve not only disaster preparedness planning or the introduction of new crops but also the capacity to identify problems created by climate change, seek solutions to them, and implement those solutions in a fair, efficient, and sustainable manner.

The Concept of "Institution"

The term "institution" as used in the social sciences generally refers to all those means that hold society together, and is formally defined as "specific or special clusters of norms and relationships that channel behaviour so as to meet some human, physical, psychological, or social need such as consumption, governance and protection, primordial bonding and human meaning, human faith, and socialization and learning" (Buttel 1997, 40). Similarly, Homer-Dixon (1999, 213) adopts the idea of institutions as "the rules of the game in a society or, more formally, [as] the humanly devised constraints that shape human interaction." Institutions are defined as "stable and predictable arrangements" for the coordination of human interaction (Ferrante 2003, 5); as "social practices" that involve power and authority (Ishwaran 1986, 247); or as "sets of norms, values, and beliefs, developed to resolve" recurring social problems (Hagedorn 1994, 367).

The variety of definitions permits a multiplicity of applications, allowing a wide spectrum of analytical possibilities, ranging from organizations to hegemonic discourses, from highly formalized settings to informal arrangements. Thus, use of the term "institution" has been shaped and explained from a variety of perspectives, each with a different explanation of the logic that motivates institutions, their origins and changes, their relationships to history, level of analysis, and individuals (Jordan and O'Riordan 1999).

Multiple definitions and perceptions surrounding the term "institution" create challenges in its use for studies of adaptation to climate change. The World Bank (2003) adopts a pragmatic approach, defining institutions as "the rules, organizations, and social norms that facilitate coordination of human action." The important advantage of this definition is that it includes organizations as part of the definition and facilitates operationalizing the term "institution" for research purposes. Organizations link people and major social institutions, thereby constituting a more concrete representation of an institution. Newman (2004) follows a similar approach, defining institutions as "stable sets of statuses, roles, groups, and organizations that provide the foundation for behavior in certain major areas of life" (302). Thus, an institution is an underlying, durable pattern of rules and behaviours, and an organization is its changeable manifestation. All organizations have a fundamental role in organizing society and its relationships with the environment. Formal public agencies are central to any discussion about institutional adaptive capacity, since they have a purposeful mandate, a degree of longevity, social acceptance, and a legal basis.

Formal organizations are not the only institutionalized settings that exist in society, however. Institutions also take the form of less formalized settings where there are no socially recognized organizational structures and specific purposes attached (Haas et al. 1993, 5). Communities and households are good examples of informal settings that have the capacity to define the parameters of the behaviour of their members and the nature of their relationships, in spite of not having the highly formalized nature of bureaucratic organizations. They are formed not just by groups of people living in the same area or under the same roof but also by symbols, discourses, norms, and all those elements that make organized everyday life possible. Like many other human settings, however, they function within institutional systems that link those settings with the larger society. These institutional systems pervade the lives of the community members by imposing a body of regulations, rules, processes, and resources that may either support or conflict with the capacities of those communities and households, impositions that are carried out by organizational structures such as public agencies (see Alcorn and Toledo 2000, 218).

Both formal and informal institutional actors exist within a larger societal context that imposes dynamics upon the ways in which households, communities, and public institutions operate. Thus, an assessment of institutional adaptive capacities must not only consider the capabilities of these formal and informal institutional actors but also recognize national institutional dynamics, such as the federal system, which imposes different functions and responsibilities on the central and provincial governments. Such dynamics influence and shape the organization, operation, and functions of the institutional actors.

Public Institutions and Adaptive Capacity

The adaptive capacity of a public institution should be understood not only as an ability to reduce its own exposure to climate risks but also as the ability to perform functions that facilitate the adaptive capacity of its constituencies. In his discussion of "institutional capacity" and climate policy, Willems (2004, 8) describes the nature of institutional capacity, arguing that it is the "ability (of a certain country) to mobilize and/or adapt its institutions to address a policy issue, as climate change."

While adaptive capacity is linked to access to resources, such access by itself is insufficient. Adaptive capacity is also more than a straightforward technical issue. (See the discussion in Chapter 4.) The development and implementation of technological measures by public institutions could be an important contribution to reduce the vulnerability of different social groups, but Adger (2003, 30) reminds us that these technological solutions could be problematic for two reasons. First, they tend to have a socially differentiated impact, benefiting some sectors of society at the expense of others, a factor that could multiply the negative consequences of climate change, producing "double losers" and "double winners." Second, their contribution to adaptation to climatic variability within the existing coping range could be high, but, as noted in Chapter 4, this range may change in a radical way under new parameters created by climate change.

Climate change could have serious impacts on the availability of resources, the viability of human settings, the livelihood of sectors of the population, and, in the long term, the social processes that characterize the relationships between the civil society and the state. In these terms, climate change affects the paths that promote sustainable development. The adaptive capacity of society is influenced by a multiplicity of factors reflected in the economy, the state, the civil society, and culture. Such factors involve technology, assets, capital resources, human and social capital, scientific knowledge, and institutional capacities such as effective social networks and flexible and innovative organizations. In a sustainable society, this multiplicity of factors would be organized in a cohesive and coherent manner that serve to increase the adaptive capacities of society. In other words, sustainability requires not a myriad of unrelated adaptive measures but a *structured* adaptive capacity.

Public institutions require flexibility to deal with the unanticipated conditions that may result from the impacts of climate change. Their role includes implementing an enabling environment that strengthens civil society to deal successfully with the challenges of climate change. Impacts on the city of New Orleans during Hurricane Katrina in 2005 serve as an example of the importance of public institutions in ensuring that residents can deal with major weather events. As Smit and Pilifosova (2003, 22) argue, "adaptation is less about identifying and implementing specific climate change adaptation measures and more about strengthening an ongoing process where resources are available to identify vulnerabilities and employ adaptive strategies." To be successful, adaptive capacity must allow for the identification and resolution of peoples' problems and the satisfaction of their needs in a fair, efficient, and sustainable manner. In this context, the adaptive capacity of public institutions is related to their ability to anticipate problems and to manage risks and challenges in a way that balances social, economic, and natural interests.

Dealing with the complexities of the impacts of climate change requires a policy and management approach based on institutional arrangements that are necessarily different from those fashioned around traditional policy problems. Climate change impacts are not limited to specific sectors, such as the economy or the environment. Thus, approaches and arrangements to address climate change impacts should act across traditional sectors, issues, and political boundaries, and address complexity and uncertainty.

There are at least five key principles informing an institutional approach for addressing climate change impacts and adaptive capacity (Alfaro 2004):

- *persistency,* where political efforts are maintained over time, enabling the accumulation of learning experience
- *purposefulness,* where political efforts are supported by stated principles and goals
- *information-richness and sensitivity,* where the best information is sought and made widely available to sustain the political efforts
- *inclusiveness,* where the full range of stakeholders are involved in policy formulation and in management
- *flexibility,* where there is a preparedness to experiment, preventing persistency and purposefulness from becoming rigidity.

Institutional arrangements that support these principles are systemically related to input, processing, and output factors. Thus, there needs to be openness of the political system to identify problems and issues in the civil society, combined with an ability to seek solutions to those problems and a capacity to implement solutions. Several authors describe the capacities that institutions require to deal with challenges such as resource scarcities (Homer-Dixon 1999) and the challenges of sustainable development (World Bank 2003). They identify many of the components that should define the adaptive capacities of public institutions.

On the *input side*, relative to the rural agricultural areas of the SSRB, it is essential for institutions to have knowledge of the current physical and social vulnerabilities existing in the agricultural sector, and to identify early the impacts of climate change upon natural and social resources. An institution must be "sensitive to early signs of problems" (World Bank 2003, 185-86) related to the impacts of climate change. The existence of appropriate information systems that allow for the gathering and evaluation of information able to support decision-making processes (referred to by Homer-Dixon as "instrumental rationality") is a central factor in fostering such sensitivity. The issue is not only the capacity to collect information but also the quality of the collected data in terms of identifying local problems and issues and the needs of different social groups, as well as the ability of the institutions to "return" this data to different constituencies.

On the *processing side,* the identification of vulnerabilities imposes a fundamental task upon public institutions: to resolve the identified problems in ways that balance the interests of a diversity of stakeholders. The capacity to resolve problems requires arrangements that are "internal" to institutional actors, such as the existence of proper resources in the institutions and their ability to link to other institutions in order to coordinate the solution of problems. Some of these arrangements include:

- avoiding policy measures or programs that may favour specific stakeholders to the detriment of others, and the consideration of the diversity of interests during the process of reducing identified vulnerabilities. For example, the World Bank (2003, 187) emphasizes two elements in this process of balancing interests: getting everybody represented in the decision-making process and facilitating the negotiation process.
- institutional features such as transparency, performance reporting, and accountability (e.g., Office of the Auditor General 2004; Stratos 2003) that promote fairness and provide the opportunity for self-evaluation (World Bank 2003, 187).
- forums and networks of negotiations during the process of finding the best solution to the vulnerabilities identified in the input side (World Bank 2003).
- availability of resources, human capital, and fiscal resources within the institutions (Homer-Dixon 1999).
- coordination among different public institutions and their capacity to agree and act on shared bases, objectives, and methods.

- limiting institutional barriers, such as management practices that affect the decision-making processes (e.g., the existence of highly centralized structures of power within institutions).
- recognition of climate change adaptation options as a viable strategy in the mandates and decision-making process of the institutions.

On the *output side,* the adaptive capacity of institutions reflects their ability to implement solutions. This feature will vary according to how well they can communicate their decisions and the implementation procedures to those they serve. Also important for public institutions is their success in promoting capacity building and problem solving within civil society. Examples include fostering of social capital and networks for mutual support within rural communities. In addition, public institutions display more adaptive capacity if they can monitor how the solutions have been worked through and evaluate their degree of success.

In these terms, the role of public institutions in the development of a climate change adaptive capacity – or institutional adaptive capacity – is most clearly reflected in governance. Governance focuses on relationships between civil society and the state, a relationship where public institutions play a fundamental role in reducing the vulnerability of stakeholders (Hall 2005). In the context of climate change, governance involves the allocation and distribution of resources, not only natural resources but also the economic, social, and political resources that are fundamental for coping with new climatic conditions.

The process of developing successful adaptive capacity in which governance plays a fundamental role involves the organization of material and human resources in order to resolve questions of sustainability: what should be sustained, how to do it, and for what purposes. In these terms, it is a political process oriented towards organizing the distribution of society's resources in different ways, that is, ranging from a neoliberal, free market society to a highly centralized society. The specific form of governance depends a great deal upon a variety of discourses – value frameworks, paradigms, and models – that are articulated by the many and varied constituents making up the social and political spectrum of society. These discourses are important because they not only define the nature of the problem but also frame the possible solutions. Thus, public institutions' roles in the development of an adaptive capacity reflect different core values and the political and cultural paradigms upon which they are explicitly or implicitly founded.

Adaptive Capacity in the South Saskatchewan River Basin

The SSRB stretches from the Rocky Mountains across southern Alberta and Saskatchewan, covering an area of 420,000 square kilometres with an estimated population of 1.5 million. The basin is divided into five major watersheds: Bow, Oldman, Red Deer, South Saskatchewan (Alberta) and South Saskatchewan (Saskatchewan). Approximately 65 percent of the basin population lives in major urban centres, mainly Calgary, Lethbridge, Medicine Hat, Swift Current, and Saskatoon, while the rural population is spread among approximately 225 towns and villages and their surrounding areas (Sobool and Kulshreshtha 2003). The study area depends economically on crop and beef production, as well as on the food processing industry, petrochemical industry, hydropower generation, and mining, mainly potash and oil and gas (Lac 2004).

Land use is primarily large- and medium-scale agriculture. The area includes the largest number of farms in Canada with 51-80 percent of their area cultivated to wheat (Lemmon and Warren 2004) and is cropped mostly with only fifteen field crops (grain, oilseeds, and pulses) and a few forage crops (Canadian Council of Ecological Areas 2004). Livestock production is also a main agricultural activity, with large areas left for pasture. There are numerous dams, reservoirs, diversions, and irrigation projects. In southern Alberta, thirteen irrigation districts divert about 2.3 billion cubic metres (1.8 million acre-feet) of water to irrigate about 500,000 hectares (1.2 million acres) of land. Approximately 120,000 hectares (300,000 acres) of land are irrigated by twenty-five irrigation districts throughout southern Saskatchewan. Besides supplying water for irrigation, the basin is used for recreation and hydro-electricity, and is the principal source of household water for 45 percent of Saskatchewan's population, including cities such as Regina that are outside the geographical area covered by the basin.

A fundamental problem for the agricultural sector in the basin is the availability of water, which is a serious problem in the context of the expected climate change impacts for the area related to increasing evapotranspiration and aridity. The current availability of water is already precarious in areas of the basin. The allocation of irrigation licences has reached its limit in southern Alberta, although irrigation farming could still be improved through better management and technologies. These improvements are problematic, however, since there is less water return to rivers and the quality of the water could be seriously affected (Lalonde and Corbett 2004). The demand for water in the basin is expected to increase significantly as a result of the expansion of the economy in southern Alberta. By 2046, the demand for non-irrigation consumptive use is expected to be between 63 and 132 percent higher than today, mainly as a result of the expansion of industry and of cities such as Calgary (Lalonde and Corbett 2004). This increasing demand is problematic in the context of the expected impacts of climate change upon the water resources of the basin.

A number of federal government organizations have specific roles related to water management issues in the basin. These institutions include Environment Canada, Agriculture and Agri-Food Canada, Health Canada, Parks

Canada, Natural Resources Canada, and the National Water Resource Institute, among others. The direct management of the SSRB water resources, however, involves integrated planning from the three provinces: Alberta, Saskatchewan, and Manitoba. In 1969, these provinces created a sharing system articulated in the Master Agreement on Apportionment, which continues to guide Prairie Provinces Water Board activities to this day. Under the agreement, Alberta and Saskatchewan are each not to exceed the use of 50 percent (net depletion) of the natural flow within their respective boundaries. Furthermore, each province should not exceed the use of 50 percent (net depletion) of the flow entering the province (Lac 2004).

The provincial governments are responsible for the management of their water resources to meet their commitment to the Master Agreement on Apportionment and to ensure water availability and water quality for all non-irrigation consumptive users. The provincial governments of Alberta and Saskatchewan are also responsible for monitoring water level and streamflow. Two main provincial agencies are in charge of these tasks: Alberta Environment and the Saskatchewan Watershed Authority. These agencies have developed water management strategy frameworks (Water for Life in Alberta, and the Saskatchewan Water Framework). Alberta is also developing a multi-phase water management plan for water use in the SSRB that involves input from four multi-sector stakeholder Basin Advisory Committees and the general public. The provinces establish and update drinking water quality objectives, but most municipalities control their own water systems. Municipal governments operate water and wastewater utilities, and have primary responsibility for providing safe drinking water to households. In addition, federal agencies such as the Prairie Farm Rehabilitation Administration (PFRA), a branch of Agriculture and Agri-Food Canada, also play central roles in supporting agricultural producers' management of natural resources. The PFRA offers programs, services, and technical assistance (and sometimes financial assistance) in many areas related to agriculture, including water supply development, wastewater treatment, irrigation, and soil and water conservation.

The foregoing description of the public institutional system relevant to water resources in the SSRB indicates the existence of many institutions that work at a variety of levels implementing a wide variety of programs. A number of civil society organizations also embrace a broad mandate on sustainability issues relevant to the SSRB. The Canada West Foundation, for example, is a charitable organization that conducts and communicates economic and public policy research of importance to the four western provinces. Its "natural capital" project is intended to highlight the importance of natural capital, including water resources, in sustainability policy discussions.

In order to meet common needs across a large geographic area, organizations have formed larger associations to represent their collective interests. For example, the Partners for the Saskatchewan River Basin (PFSRB), based in Saskatoon, promotes stewardship and education across the entire basin of more than 3 million people who depend upon the North Saskatchewan, Red Deer, Oldman, Bow, Highwood, South Saskatchewan, Battle, Saskatchewan, St. Mary, and Carrot Rivers. In order to accomplish this mission, the PFSRB develops public awareness and education tools, facilitates partnerships and networks of organizations that cross political and sectoral boundaries, and designs and implements stewardship action projects. Other associations focus on promoting networks throughout a province. The Saskatchewan Network of Watershed Stewards (SNOWS) is an example of a partnership involving provincial, federal, and non-governmental organizations, designed to coordinate and support watershed stewardship programs in that province.

Some associations focus on particular geographic areas within the larger water basin. For example, the Bow River Basin Council, originally established as a water quality association, operates as an arms-length advisory council to the government of Alberta, with 120 members representing the commercial/industrial sector, individuals, irrigation licensees, non-profit, academic, municipal, and administrative/regulator bodies, and First Nations interests in the basin. The council was very involved in developing the Alberta water strategy, state-of-the-river reports, and the SSRB water management plan. There are several other groups in Alberta similar to the Bow River Basin Council, including the Chestermere Watershed Committee, the Nose Creek Watershed Partnership, the Iron Creek Watershed Improvement Society, and the Oldman Water Quality Initiative.

Some associations focus on smaller watershed areas, such as the Swift Current Creek Watershed Stewards or the Turtle Lake Watershed Partnership in Saskatchewan, while others have a broader focus on sustainability issues at a regional scale. For example, Alberta's Prairie Conservation Forum is a partnership among government and non-governmental organizations that allows members to discuss a wide array of sustainability topics related to the Prairies, including climate change and water conservation. The Saskatchewan Prairie Conservation Action Plan (PCAP) partnership brings together over twenty-five industry, agricultural, government, non-governmental, and academic representatives to focus on native prairie conservation, and includes education programs in rural areas, such as the "Cows, Fish, Cattle Dogs and Kids Game Show," which fosters an awareness of the interrelationship among all ecosystem elements. In another collaborative initiative, provincial and federal government agencies are collaborating on a southern Alberta sustainability strategy that aims to assess the socio-economic vitality of southern Alberta without significantly impacting the environment. Other associations focus on particular aspects of water management, such as the Alberta Irrigation Projects Association and the Saskatchewan Irrigation Projects Association, which provide irrigators with the opportunity to meet and work together. Other initiatives are established for specific periods of time, such as the Wonder of Water program. Initiated in 2003 for a two-year period, that program promotes the establishment of longer-term partnerships focusing on water conservation.

Research networks have been developed to facilitate collaborative work among scholars and others. The Water Institute for Semiarid Ecosystems (WISE), based at the University of Lethbridge, is a consortium of scientists and people from industry, environment, agriculture, and the irrigation districts. The Prairie Adaptation Research Collaborative (PARC), based in Regina, fosters a wide range of research related to adaptation issues on the Prairies.

These examples of programs and initiatives in the SSRB reflect the critical importance of water as a resource to meet a multitude of needs throughout the basin. The increasing recognition of impacts from climate change in an area that has a long history of adapting to water shortages has contributed to the development of government partnership programs focusing directly on climate change. For example, Alberta Environment has a project to assess awareness of adaptation issues within its own organization, and is developing a climate adaptation strategy. At the same time, it participates in an interdepartmental climate change working group. This openness to institutional partnership has also spread among other public and civil society institutions that participate in the management of water resources.

The existence of this network of government and civil society organizations in the SSRB and their willingness to establish partnerships create significant potential for the development of an institutional climate change adaptive capacity. The network's predisposition to partner with other organizations could facilitate the production and dissemination of knowledge of the current physical and social vulnerabilities existing in the basin and the early identification of the impacts of climate change upon natural and social resources, as well as provide forums for facilitating negotiation processes oriented towards a more rational and fair use of the water resources. Moreover, the network could also facilitate coordination among different public institutions and between public and civil society organizations, as well as their capacity to agree and act on shared bases, objectives, and methods in order to face the challenge of climate change.

It is difficult to predict the types of discourses that might prevail in the process of developing this institutional adaptive capacity. One of the few empirical studies in this area is a relatively recent study of the perceptions and attitudes of members of the natural resource policy community in the

Prairie provinces, a community consisting of members of any institution that has influence over the formulation of policies (Wellstead et al. 2002). The study shows that droughts, water supply, and climate change are among the most important policy concerns, ranking at the top of a list of fifteen current natural resource issues. The study classified the values, or core beliefs, of the most important members of the policy community – the representatives of industry, government, universities, and environmental organizations – in two domains: an economic domain (with an emphasis on support for private property rights and free market economic expansion) and an environmental domain (with an emphasis on nature and limits to growth). The results show that industry and government members of the policy community are more inclined to support industry interests than ecological concerns, as opposed to university and environmental movement representatives, who are predisposed towards ecological values (Wellstead et al. 2002). This value system could have the potential to influence the development of an adaptive capacity oriented towards ensuring economic sustainability over a more balanced sustainable development.

This institutional setting has already demonstrated its capability to foster the development of an adaptive capacity in response to current climate-related vulnerabilities. The establishment of the PFRA is an example of an institutional response to a set of climate-related conditions. As noted in Chapter 9, the PFRA was founded in response to the drought of the 1930s, as a federal government initiative to assist agricultural communities in the Prairies to develop agricultural processes and techniques that would lessen the vulnerability of the farming and ranching communities to climate-related stresses. Over its many years of service, it has operated a variety of soil and water assistance programs, including extension services, infrastructure grant programs, and many others, becoming an important source of support for both agricultural producers and rural communities. For example, it played a central role in ensuring the sustainability of Cabri, a rural community in one of driest areas in southern Saskatchewan. In the early days, water was a precious resource in the area of Cabri, being hauled for miles for irrigation and consumption. In the early 1950s, the PFRA was asked to provide a solution that would secure a stable source of drinking water for the town population. The result was the building of a water reservoir by the PFRA, a reservoir that is still used by the locals.

At a less formal institutional level, there are also examples of adaptive capacity within rural communities. A recent study of the climate change adaptive measures of Alberta agricultural producers shows a significant number of past and current adaptive strategies (Stroh Consulting 2005). Measures include changing the location of their operation, using shelterbelts and bushes to conserve moisture, crop insurance, changing crop types and varieties, reduced or zero tillage, and others. In terms of future strategies, farmers mention the adoption of organic farming, enhancement of crop insurance, and – especially relevant to the development of an institutional adaptive capacity – the need for governments to increase their role in the development of adaptive capacities among farmers, particularly in the area of education and dissemination of information (Stroh Consulting 2005). At a collective level, Adger (2003) has emphasized the role of social capital as an adaptive capacity, arguing that this form of capital is an important element in coping with climatic variability and hazard. This form of capital is also prevalent within the multitude of rural communities that exist in the SSRB, and is an important mechanism used by people to cope with drastic changes, such as the structural transformation of agriculture, and the problems of everyday life (Diaz and Nelson 2005).

The existing institutional setting also has its limitations, as was demonstrated in the case of the drought of 2001 and 2002. That drought, which could be interpreted as a harbinger of a typical natural hazard under future climate scenarios, had repercussions that went far beyond agricultural production, affecting recreation, tourism, health, the supply of electric power, transportation, and forestry. A recent study of the impacts of the drought indicates that in spite of the government response and the safety net programs, "the wide array of adaptation measures, including government programs, could not cope with the immensity of the losses," especially in the west (Wheaton et al. 2005, 23).

These limitations appear to be related to a lack of preparation to deal with climate-related problems and to the fact that coordination among institutions is very limited. The study of Wittrock and colleagues (2001) shows that public institutions lack the necessary awareness and preparation to deal with many climate-related problems. There is the need, the authors argue, for more information about climate impacts and adaptation, and increased research to facilitate the decision-making process around issues that are affected by climate (Wittrock et al. 2001). The study of water resources and climate change in the city of Regina shows the same limitation (Social Dimensions of Climate Change Working Group 2005). The different institutions that participated in the study were aware of climate change and its potential impacts upon water resources. This awareness, however, had not translated into a significant integration of climate change issues into the institutional agenda and organization. In the best of the cases, it had been limited to the assignment of resources to promote and monitor the reduction of greenhouse gas emissions.

Not surprisingly, the existence of a wide array of institutions in the SSRB engaged in numerous activities across a variety of temporal and spatial scales with varying degrees of focus on water management and conservation and climate change issues is characterized by problems of coordination among all of those institutions or activities. In the study on climate change and water resources in the city of Regina, the issue of institutional coordination with regard to water quality was raised by participants in a focus group. Participants were very critical of the capacity of public institutions to gather standard information about water quality. This information, it was argued, is collected by different institutions and for different purposes as a result of the lack of coordination among the different levels of government and institutional squabbles about who is in charge of particular issues. Such a concern has been echoed in several informal interviews with representatives of public institutions, who claim that it is one of the most fundamental problems of government organizations. It appears that institutions tend to focus on areas relevant to their mandates, which are mostly defined by a perceived set of problems, the clients they serve, and sources of funding for their services. Also, institutions operate at specific levels or scales of influence, whether they are local municipal governments, national non-governmental associations, or local business associations.

We have argued that the social impacts of climate change are likely to be heterogeneous given the variable impact of climate change among different socio-economic sectors. Formal and informal institutions should be structured in a manner that facilitates the development and implementation of comprehensive support mechanisms that improve the capacity of different sectors to adapt to climate change. Such an institutional structure would be characterized by transparency in decision making, performance reporting, and accountability, which promotes fairness and provides the opportunity for self-evaluation. A fundamental focus, particularly of public institutions, should be the *anticipation* of problems so that risks and challenges can be managed in a way that balances social, economic, and natural interests. The ability to anticipate requires knowledge of current physical and social vulnerabilities. Thus, institutions should include a focus on optimizing the resources available in order to identify vulnerabilities and employ adaptive strategies. Our study of the SSRB reflects the critical importance of the adaptive capacity of institutions in the management of water as a resource in order to meet a multitude of needs throughout the basin.

Notes

1 Material in this chapter is linked in part to the project "Institutional Adaptation to Climate Change," supported by the Major Collaboration Research Initiatives (MCRI) Program of the Social Sciences and Humanities Research Council of Canada. It incorporates several of the arguments developed in a working paper for that project entitled "Institutions and Adaptive Capacity to Climate Change" by H. Diaz and A. Rojas (http://www.parc.ca/mcri).

11

The Perception of Risk to Agriculture and Climatic Variability in Québec: Implications for Farmer Adaptation to Climatic Variability and Change

Christopher Bryant, Bhawan Singh, and Pierre André

This chapter considers how the perception of risk from climatic conditions figures in adaptive decisions in Québec agriculture, and offers conclusions in relation to policy issues and development. The review recognizes the various factors that influence human decision making, thereby establishing a context for the adaptation research. Like impact-based approaches, the empirical analysis presented in this chapter employs climate change scenarios. These projections are presented to farmers and elicit responses regarding the likely consequences for managing their farming operations. Similar to process-based studies, this analysis uses farmer interviews to gain insights into risks for agriculture.

There has been substantial interest in climate change and Québec agriculture since the early 1990s. We have been involved in much of the related research and, as an introduction to more recent findings, offer an overview of climate change and agricultural research in Québec. This review serves as a recap and a reminder of the points outlined in Chapter 2, where the evolution of different research perspectives and the factors giving rise to them are discussed. As well, we identify at a number of points how research in Québec is related to research in agricultural adaptation to climate change and variability undertaken by researchers in other parts of Canada. In this evolutionary approach, the intention is to compare the different approaches to the research questions that have evolved over time. We begin with such a discussion, stressing both the methodological approaches used as well as how the research questions have evolved. An important ingredient in the methodologies is the relationship between the information sources used and the evolving research questions. The perspectives that emerge from this research are discussed next, particularly in the context of farmers' risk management strategies. The structure and early findings of the current research program are followed by some considerations for policy.

Agricultural Adaptation and Climate Change Research in Québec

As noted in Chapter 2, where the impact-based approach is reviewed, much of the earlier work on impacts of climate change and variability on Québec agriculture focused on the generation of future climate scenarios. These permitted the calculation of climatic parameters and agroclimatic variables that were then incorporated into crop models for different crop types and in different regions under study, to derive crop yield changes in response to climate change and variability. It was thus possible to compare future yields under different climate scenarios with current yields, giving "impacts" in terms of changes in yields (Bryant et al. 2000). The yields for different crop types could then be compared, and implications for change in agricultural land use were derived directly from these model outputs. This was done in Québec as elsewhere in North America (Mearns et al. 1992; Rosenberg et al. 1992; Semenov and Porter 1995; Singh and Stewart 1991).

The need to incorporate the "human factor" in climate change adaptation research resulted in a comparable change in orientation that included human agency with the biophysical impact-based approaches (André et al. 1996; Singh et al. 1996; Singh et al. 1998). From there, the issue of the adaptation of agriculture to climate change and variability (Bryant et al. 1997) was highlighted, followed by effort directed towards understanding the capacity for adaptation of different farmers and farming systems (e.g., Bryant and André 2003).

The research questions in more recent studies (as for several other research groups in Canada) have focused on how farmers perceived the risks from climate change and variability in the context of their decision environment, as well as on the types of strategies farmers felt might be appropriate. Farmers thus provided the key, mediating link between the changes in the biophysical environment (climate change and variability as they relate to agricultural production – the "exposure factor") and actual changes in agricultural crop mixes and practices, thereby raising explicitly the issues of adaptation and vulnerability. Subsequently, in response to results from the research that consistently identified climate change and variability as being of interest to farmers but not of major concern, the research thrust began to investigate how farm strategies might reduce or increase farm vulnerability to climate change and variability, regardless of whether the initial impetus for the strategies was climate variability (André and Bryant 2001). Probably the major contribution from the work undertaken in Québec up to the late 1990s and very early in the new century was the emphasis on the need to ground understanding of farm adaptation and farm vulnerability in the realities of farming and farmer decision making.

Increasingly, the research focus has broadened to include policy implications, both in relation to the principal government agencies concerned and other players who have a stake in farming and who can influence directions through their own actions, programs, and policies, such as the Union des producteurs agricoles du Québec (UPA) and La Financière agricole du Québec (FAQ). At the same time, while a significant effort is still being made to integrate changes in the biophysical environment with the human agency factor, there is a more systematic effort to link the implications of possible future changes with the experiences of farmers in dealing with problematic or extreme climatic conditions in the past. That is, in keeping with process-based approaches (see Chapter 4), current research tries to understand farmers' adaptive capacity for managing recent stresses and change so that some idea of their capacity for dealing with future climate change can be gained.

The Information Base

As the climate change adaptation research preoccupations and questions evolved for Quebec agriculture, so did the methodologies[1] and information sources that the studies relied upon. In the early to mid-1990s, research that attempted to integrate changes in the biophysical environment and human agency generated future climate scenarios under different assumptions, calculating critical agroclimatic indicators for these scenarios and then using these agroclimatic indicators to produce future patterns of yields for selected crops in specific regions. These results were then presented to farmers in order to obtain their responses. Historical climate data therefore constituted critical input into this research, as did the various crop models (e.g., FAO-modified model; DSSAT 3.5) (El Mayaar et al. 1997; Singh et al. 1995, 1998; Stewart 1983).

Interviews were conducted with samples of farmers in the different regions studied, with the aim of determining their reactions to the issue of climate change and how they had dealt with unfavourable climatic conditions in the past. In addition, groups of farmers and other professionals involved in agriculture were brought together in workshops (or focus groups) in several regions. Early efforts centred on the South-West Montréal region as well as the Québec region (Lotbinière and the Île-de-Orléans). The groups were confronted with future climate scenarios for their region, as well as with the results of the crop modelling exercises for selected crops (e.g., grain corn and selected vegetable crops in the South-West Montréal region). Participants were asked a variety of questions: How did they react to the scenarios (climatic conditions, agroclimatic indicator values, and postulated crop yields under the different scenarios)? Would such conditions and "results" be of concern to them? How would they respond? What sorts of strategies might they envisage as a way to deal with the new conditions?

When the research efforts also began to include a greater emphasis on adaptive capacity and vulnerability at the farm level, the research methodology shifted away from the modelling of future climate scenarios to a focus on farm interviews followed by a panel discussion format involving representatives of stakeholder groups (e.g., the UPA and the FAQ). In an attempt to identify the important indicators of adaptive capacity at the farm level, the panel was presented with several case histories of actual farms and their farming families (who remained anonymous), for which the evolution of crop mixes and farming practices – including the type and quantity of labour used – were presented. The panel then identified the critical indicators for distinguishing farms based on their demonstrated capacity to adapt to change (change generally, not just climate change and variability). The results were then used to analyze data from the sample of farm operators who had been interviewed (André and Bryant 2001; Bryant and André 2003).

In the most recent research efforts into understanding farmers' adaptation and vulnerability to climate change and variability in Québec, a combination of all of the above methodologies is being used. Besides the information sources identified above, two relatively new sources of information are being employed. First, given the identified need to ground research (and policy) in the reality of farming and the farmer's decision-making environment (C-CIARN Agriculture 2004), researchers are making extensive use of crop insurance claims and participation records from La Financière agricole du Québec (these are reviewed in the "Current Research on Farmer Adaptation to Climate Change and Variability" section of this chapter) as a means of identifying temporal and spatial patterns of peaks in agricultural crop losses. Second, farm-level modelling will be developed in the 2005-07 period to assess the costs and benefits of adapting or not adapting to particular scenarios of future climatic conditions. Thus, researchers will be in a position to generate and assess adaptation and non-adaptation scenarios by using downscaled climate scenarios (coupled with crop models) to provide the climatological and yield impact information, in combination with the evidence from farm interviews and focus groups.

Farmer Adaptation: Perspectives from Research

Given the evolving research questions and preoccupations, and the concomitant shifts in methodologies and information sources, what has been learned so far from the research on farm adaptation to climate change and variability in Québec?

Our comments are structured around the themes of farmers' decision-making environment, the context in which farmers must function, and their perceptions of climate change and variability. First, echoing work on agricultural change around cities and elsewhere over the last three decades (e.g., Bryant and Johnston 1992), the research undertaken in the early 1990s emphasized the complexity of farmer decision-making environments. Farmers make decisions in relation to a multitude of stressors and opportunities that appear in their decision-making environment, such as changes in international and interregional competition, political decisions on support systems at home and abroad (which affect competitors), and technological change. (See Chapter 8 for more details.) Some stressors require an immediate decision (even a decision to take no action), while others are longer-term in nature, perhaps requiring immediate decisions but also requiring an ongoing process of assessment and adjustment. How farmers perceive the relative importance of different stressors and opportunities can therefore be expected to have a material impact on what decisions they actually make.

Second, in view of this complex decision-making environment, the context in which farmers operate is important because it contains the parameters, which contribute to the setting of ranges for farmer decisions on strategies and investments. Two aspects of this broader context are worth noting, both of which have had an impact on the structure of the most recent research interests, described in more detail later in this chapter:

- *the role of additional factors that are relatively specific to particular agricultural sectors.* These include particular market forces and patterns; appropriate technologies and their evolution; and patterns of government intervention, such as decisions affecting international trade and supply management systems, which have been particularly important in the Canadian and Québec contexts (for example, in dairy production).
- *the role of other players whose actions, policies, and programs contribute to defining at least part of the farmer's decision-making environment.* Included among such players are ministries of agriculture, crop insurance programs, and professional organizations for farmers. Other players, such as consumer advocacy groups and environmental protection associations and groups, have a more indirect impact on the farmer's decision-making environment.

The early research in Québec, where farmers and related professionals were shown scenarios of climate change and associated impacts on crop yields, revealed that farmers were indeed interested in the topic of climate change and variability. This finding is also supported by research from southern Ontario (see Chapter 13) and BC (see Chapter 12).

In the early 1990s, however, apart from a certain level of skepticism, climate change and variability tended to be seen more as a curiosity and not something that farmers felt they had to deal with immediately. Some of the farmers encountered did make the connection between past experiences with extreme climate events and asked questions about how these might relate to future patterns of change in climatic conditions. Farmers in such encounters also provided valuable information on the nature of the agroclimatic indicators that were important to them – they emphasized the importance of variability in conditions at specific points in the growing season rather than average precipitation levels or temperatures. This information was used to inform the biophysical modelling component of the research effort and the crop modelling exercises. Furthermore, information received from farmers about the types of strategies they have adopted in the past and how these might relate to coping with and adapting to changing climatic conditions contributed to the broader Canadian literature on adaptation strategies available to farmers (e.g., Dolan et al. 2001; Smit and Skinner 2002). Such strategies also included crop insurance programs, although many farmers thought of these as simply short-term measures that would help them deal with short- to medium-term conditions of uncertainty in relation to climatic conditions.

Recognition of the complexity of farmers' decision-making environment, already seen for several decades in other research domains dealing with agriculture, gave new impetus to the research effort in the late 1990s. The fact that farmers are exposed to multiple stressors raises questions about their investment and management strategies in general, as pointed out in Chapters 8 and 9. In fact, many adaptive responses can be seen as reducing farm vulnerability to climate change and variability even though they are strategies that have been adopted primarily in response to other stressors and opportunities (e.g., farm diversification stimulated by market changes and income considerations may also contribute to reducing vulnerability to certain climatic conditions for specific crops) (André and Bryant 2001).

It is important to note, however, that a number of strategies adopted by farmers do help them cope with particular climatic conditions and events. Farmers interviewed during several of the research projects often voiced the opinion that they already possessed a toolkit that would enable them to cope with the sorts of changes in climatic and weather conditions associated with the various scenarios they had been presented with. Those that stood out are water management strategies, crop choice and diversification, irrigation and drainage strategies, participation in crop insurance programs, and income stabilization programs.

This same research thrust was also associated with an increasing interest in farm vulnerability, which is related to both exposure to changing climatic conditions and the adaptive capacity of farmers (and the agricultural system generally) to respond to these changing conditions. Researchers (e.g., Bryant et al. 2004) have drawn conclusions related to adaptive capacity, and identified a set of self-explanatory indicators (emphasizing farm-level conditions as opposed to broader political and economic factors) for the capacity to adapt to changing climatic and weather conditions. The indicators are:

- financial resources and capacity for debt repayment
- capacity for critical self-analysis
- training and education
- awareness and concern in terms of knowledge
- biophysical conditions (on the farm)
- adaptability of the farm itself (fixed investment, capital with limited range of alternative uses).

The cumulative research effort directed at Québec made contributions in methodological approaches for understanding the integration of climate scenarios, crop modelling, and farmers' perceptions of climate change and variability as a stressor in their decision-making environment. One perceived need that emerged from this research was related to validating indicators of adaptive capacity more systematically and furthering understanding of how patterns of farm vulnerability in relation to climate change and variability vary regionally. These gaps set the stage for a consideration of the current research program, since the concerns emerging from some of the earlier work coincided with those of a number of other stakeholders. This led to a combination of forces in a new research program that commenced late in the fall of 2004.

Current Research on Farmer Adaptation to Climate Change and Variability

Current research,[2] beginning in late fall 2004, focuses on how farmers have developed risk management strategies in relation to climatic variability over the past thirty years, and how adapting or not adapting to variability associated with climate change may affect yields, returns, and costs, using farm-level models. Various stakeholders perceived a need to ground analyses of agricultural adaptation to climate change and variability more fully in the past experiences of farmers with extreme climatic conditions or events. These issues are directly in line with process-based approaches to climate change adaptation as described in Chapter 4. Analyzing how farmers coped with and adapted to those prior events is a necessary step in understanding the costs and benefits of adapting to future climatic and weather conditions.

The current research project thus combines the following components:

- analyses of the patterns of agricultural crop losses in relation to extreme climatic conditions over the past twenty to twenty-five years, by crop type and by region, and ultimately by municipality, thus grounding the analyses in the realities of regional agro-economic systems
- analyses of the patterns of plant sensitivity to climate and soil conditions, particularly at local levels

- analyses of farmers' risk management strategies in relation to those extreme climatic conditions, including choice of crop type and cultivars, agricultural practices, and participation in crop insurance programs
- by focusing on specific types of farms, and through farm-level modelling, assessment of the costs and benefits of adapting or not adapting to specific patterns of climate change and variability.

Ultimately, the goal was to analyze and synthesize this information to provide an assessment of farm vulnerability to climate change and variability (by combining "exposure" to climate risk with adaptation strategies and adaptive capacity).

As noted throughout Part 3, it is particularly important to see farm adaptation to climate change and variability as part of farmers' risk management strategies. Farmers employ various risk management strategies to deal with the various sources of uncertainty in their decision-making and production environment. Based on earlier research, it was considered relevant to see farmers' decisions regarding risk management in context, that is, both in the context of the broader forces shaping and influencing each agricultural sector and in the context of the various other players whose actions, programs, and policies contribute to the setting of parameters in the decision-making environment. In the case of crop insurance programs in Québec, managed by FAQ, the context for decision making has four distinct aspects: (1) information provided by FAQ to farmers in terms of manuals of "good practices" for different crops in different environments; (2) the definition and revision of crop insurance program regions; (3) the setting of participation costs for farmers; and (4) FAQ's ability and willingness to intervene to help farmers reduce potential losses if negative conditions can be flagged and assessed early enough in the growing season.

Given the nature of the research question, the methodology involves using multiple sources of information whose importance varies from one phase of the project to another. In the first phase, the focus on learning from farmers' past experiences with extreme climatic conditions and events has necessitated undertaking a provincial- and regional-level analysis of farmers' participation in, and declared crop losses to, the crop insurance program. Agricultural census data were also used to "anchor" the level of participation by farmers and crop types. This was necessary because the crop insurance data do not include identification of the total base of farms or crop areas that could have participated in the crop insurance program. The analysis of the first phase has extended over a period of twenty-two years for a number of crops (e.g., grain corn, hay) and somewhat shorter periods for other crops (e.g., soy beans). In this same phase, the database provided by La Financière also contains an expert assessment of the principal causes

Figure 11.1

Evolution of grain corn yields, Québec, 1987-2001

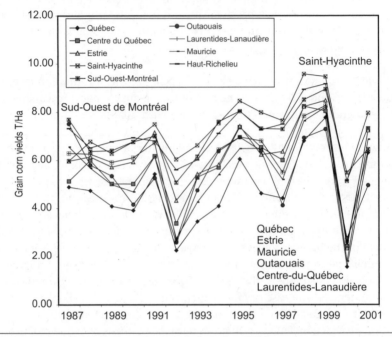

Source: Data supplied by La Financière agricole du Québec (2004).

of the claims; in addition, the annual reports, as well as intra-season reports provided by La Financière, provide more qualitative assessment summaries of the principal causes of crop losses and the regions most concerned.

As an example of the analyses, Figure 11.1 shows the evolution of grain corn yields as declared by the participants in La Financière's crop insurance program. An overall trend towards increasing yields is apparent over the 1990s, except for 1992 and to a lesser extent 1997, followed by a marked drop in yields in 2000. The year 2000 was an exceptionally bad year in terms of crop production. There are substantial regional differences, however, with the Saint Hyacinthe/South-West Montréal/Haut Richelieu areas standing out with yields consistently above the other regions, while the regions characteristically regarded as having more problematic environments for grain corn production have consistently lower yields (e.g., Québec and Outaouais).

In an associated analysis of the causes of crop losses for the period 1982-2003 as indicated in the comments provided by La Financière on each declaration, it is interesting to note that *grain corn losses were most often associated with excess rain and early and late frosts, as opposed to water deficits or drought conditions.* The causes most often associated with hay losses were again excess rain first, followed by drought conditions and early frosts. Vegetable losses, on the other hand, were associated with drought first, and then hail and excess rain, reflecting to a large extent the areas where vegetable production tends to be concentrated in Québec.

At the end of the first phase, three regions (the Montégégie southwest of Montréal, the Centre-du-Québec, and Saguenay-Lac-Saint-Jean) were selected to represent a range of agroclimatic conditions. In addition, among the regions, crops are grown that are considered particularly important. For example, grain corn is present in all three regions and is an important economic crop in Québec. In one recent year, growers experienced substantial losses (payments for grain corn losses accounted for about 75 percent of all payments under the crop insurance program in Québec in 2000). Hay is present throughout, but especially in the Saguenay-Lac-Saint-Jean region; it was considered an important crop because of its link with the all-important dairy sector in Québec. Vegetable production is more concentrated, including part of the southwest of Montréal region, and is a vital component of the food processing industry.

In the second phase, underway in the spring of 2005, the research focuses on the intra-regional patterns of crop losses within the three principal regions selected for more detailed study. Already, the patterns emerging in terms of the loss index[3] (either for all crops or for specific crops such as grain corn) provide evidence of significant intra-regional variation in all regions, but especially in the Centre-du-Québec and Saguenay-Lac-Saint-Jean regions (Figure 11.2). These intra-regional patterns in crop loss will then be aligned with analyses of climate data over the same period of time, derived from the meteorological stations in each region. The purpose of this analysis is to identify "hot spots" (i.e., areas where crop loss has been extreme) in each region where more detailed investigation will then take place.

The third phase involves working in more detail in selected "hot spots," and undertaking farm-level interviews and focus groups with farmers and other professionals involved in agriculture in these areas. This phase of the research will therefore focus on how farmers coped with and adapted to the specific extreme climatic conditions and events identified (e.g., the strategies followed, including changes in crop choices and agricultural practices), as well as how they perceived these events and conditions (e.g., an unusual event, or an event that might be part of a recurring pattern). The research team can then explore with farmers and other concerned parties how these strategies might be, and the extent to which they have been, utilized to anticipate further extreme conditions.

The fourth phase will involve using programming models on selected farm types to probe the costs of adapting and not adapting to climate change

Figure 11.2

Grain corn loss index for the three study areas, 2000

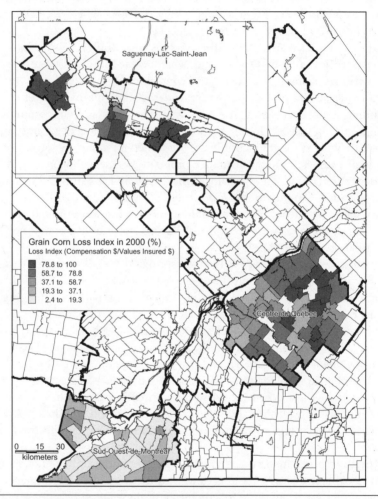

Source: Data supplied by the Financière agricole du Québec (2004).

(using scenarios generated by climate change and crop models). Finally, all of the analyses will be synthesized in order to address the question of farm vulnerability in the context of the different regions and selected crop types and farming systems across Québec.

Conclusions: Policy Issues and Development

The research effort aimed at understanding Québec agricultural adaptation to climate change and variability has been underway since the early 1990s, and in many ways includes all three approaches identified in Chapter 1. The goal of this research is to integrate the biophysical side of the equation (climate scenarios and their implications for crop yields through the use of crop modelling) with the human factor. Dessai and Hulme (2004) argue that such an integration is necessary for developing climate change policy. We concur, as shown in the rest of our conclusions.

In the Québec case, the research design has become increasingly holistic, dealing with the "exposure" component (the extreme climatic conditions and events, and the crop and farm types in different regions that make some areas, crops, and farm types more vulnerable than others) and the "adaptation" (including adaptive capacity) component of the farm vulnerability equation. Data for past climatic conditions and extremes, information on crop losses by crop type and municipality (allowing aggregation upward to any other spatial unit deemed interesting), and agricultural census data have all been combined to produce an analysis of provincial and regional conditions over a period of time. This research program involves following a line of inquiry down to the municipality level, and ultimately to the farm unit level, to provide a comprehensive picture of how farmers have adapted to extreme climatic conditions in the past two decades or so. In the final analysis, combining farm-level programming models with the information from climate scenarios will enable the research project to address the costs and benefits of adapting or not adapting to climate change and variability. The analysis thus far (spring 2005) demonstrates the reality of significant spatial variation at both the regional and intra-regional levels in relation both to crop losses and their importance and to the spatial and temporal incidence of climatic extremes associated with those losses.

There is substantial interest in the policy implications of these types of analyses and even in the preliminary results obtained so far, as climate change and variability are increasingly accepted as a reality that cannot be ignored. This is so for both governments and para-public agencies, such as crop insurance agencies. In the context of the Financière agricole du Québec, it is interesting to note that the crop insurance data are not "neutral," in the sense that the agency is relatively proactive – providing advice to farmers through "good practice" manuals, undertaking experimentation in program coverage, and being prepared to make changes in the geographic area of coverage for certain crops. In all of this, the capacity to adapt, on the part of both the farmers and the other players involved (crop insurance agencies, government ministries and agencies, professional organizations), is being identified as an important focus for future policy as part of the overall effort to reduce farm vulnerability to climate change and variability.

Notes

1. A review of methodologies and methods of collecting information on agricultural impacts and adaptation to climate change and variability research in Québec and elsewhere in Canada can be found in Brklacich et al. (1997a).

2. The research team consists of participants from the Université de Montréal (C.R. Bryant, B. Singh) and McGill University (P. Thomassin, L. Baker, C. Madramootoo). The Stakeholder Steering Committee consists of representatives from Ouranos, La Financière agricole du Québec, the Ministère de l'Agriculture, de la Pêche et de l'Alimentation du Québec, the Union des Producteurs Agricoles du Québec, and Agriculture and Agri-Food Canada. These representatives provide professional time. La Financière provides access to valuable sources of data on crop insurance, and Ouranos provides access to climate and meteorological data as well as its capability in mobilizing the various stakeholders. Funding comes from the Climate Change Impacts and Adaptation program of Natural Resources Canada (project A931) and Ouranos.

3. The loss index is calculated as the ratio of the value of claims paid out (e.g., for grain corn) to the total value insured for that crop in a given locality.

Part 4:
Process-Based Studies

One of the distinguishing features of the process-based approach is its emphasis on understanding adaptation from the viewpoint of the community of interest. First steps for inquiry are not necessarily macro-level scenarios of climate change, nor are they characterizations of the larger socio-economic and political context influencing adaptation choices. Instead, with process-based inquiry researchers engage individuals (in this case agricultural producers) to find out the nature of risks for them, the sources of risks, and how they manage the risks. The stakeholders involved in the research reveal the climatic and weather conditions that are particularly important for their operations and then consider their capacity for adapting to future conditions. Researchers work from a micro basis and consider macro-level factors from that perspective. Many of these studies yield insights about the factors that constitute the context for adaptation in agriculture.

Chapters 12 (by Suzanne Belliveau, Ben Bradshaw, and Barry Smit) and 13 (by Susanna Reid, Suzanne Belliveau, Barry Smit, and Wayne Caldwell) closely follow this farmer-based assessment and can also be classified as "vulnerability assessments." In Chapter 14, Cynthia Neudoerffer and David Waltner-Toews focus on particular ecological aspects of climate change impacts and broaden the discussion to include the related concept of resilience. In Chapter 15, Robert McLeman uses a historical community of interest badly affected by drought conditions in the 1930s to examine how access to different types of capital (social, cultural, and economic) figured in adaptation processes.

12

Comparing Apples and Grapes: Farm-Level Vulnerability to Climate Variability and Change

Suzanne Belliveau, Ben Bradshaw, and Barry Smit

Farmers have long contended with changing conditions, be they variable weather, alterations in government policy, or fluctuating commodity prices (see Chapter 8 for more details). With projected climate change, which may exacerbate current weather conditions, thereby increasing the frequency and magnitude of extreme events (McCarthy et al. 2001), the climatic context for agricultural production will likely become more complicated. While climate change is expected to present both risks and opportunities for Canadian agriculture, the implications will largely depend on farmers' ability to adapt, an ability that is currently uncertain (Brklacich et al. 1997a; Bryant et al. 2000; Lemmon and Warren 2004).

The role of adaptation as a response to climate change is gaining credibility in both policy and academic arenas, since, despite the efforts to reduce greenhouse gas emissions, some degree of climate change is inevitable (Burton et al. 2002; Parry et al. 1998). Hence, there has been a growing scholarship in the climate change field that considers or incorporates adaptation into the analysis (for a review, see Tol et al. 1998). As noted in Part 2 of this book, the bulk of these efforts in the agriculture and food area have been directed towards conventional climate change impact assessments, which estimate the biophysical impacts of future climate scenarios (mainly climatic norms) for an agricultural region (Chiotti and Johnston 1995; Feenstra et al. 1998). In these analyses, arbitrary or hypothetical adaptations, such as changing planting dates or crop types, are incorporated into the model to illustrate the potential for farmers to reduce the negative impacts or to benefit from a changed climate (e.g., Easterling et al. 1993; Kaiser et al. 1993; Mendelsohn et al. 1994; Rosenzweig and Parry 1994).

These studies have made important contributions to the identification and assessment of the costs and benefits of climate change at a broad scale. As noted in Part 1, however, their utility for identifying appropriate adaptation strategies is limited, due in part to their presumption about climatic variables relevant to agricultural decision makers and their rather simplistic treatment of adaptation processes (Burton et al. 2002; Chiotti and Johnston 1995). Climate change impact assessments typically focus on climatic variables from the global models that are presumed to be relevant to producers, such as average temperature and precipitation. It is rare for farmers to experience average conditions, however, and they are often sensitive to climatic conditions, especially variability and extremes that are infrequently included in scenarios (Smit et al. 2000a). Further, the future scenarios tend to focus upon climatic variables only, and assume that all other variables (e.g., socio-economic, political) remain unchanged. Such a portrayal does not reflect the dynamic, multifaceted decision-making environment within which farmers operate, as pointed out in other chapters of this volume. Consequently, the adaptations that are modelled tend to be those that may be applicable for the prescribed climatic stimuli alone but do not capture the other factors that may facilitate or constrain a response, and that assume that all farmers will respond in the same manner (Bryant et al. 2000; Reinsborough 2003; Schneider et al. 2000).

The focus of recent research has shifted from modelling adaptation to evaluating potential adaptation options according to criteria such as effectiveness, institutional compatibility, and flexibility (e.g., Carter et al. 1994; de Loë and Kreutzwiser 2000; Dolan et al. 2001; Smith and Lenhart 1996). Yet even these studies have limited ability to consider adaptations to multiple risks (Bradshaw et al. 2004), bringing to light two key challenges for researchers in the field of climate change impacts and adaptation. First, there is a need to improve our understanding of the likely uptake of adaptations to climate change. This requires an appreciation of farmers' ability to adapt or cope with *current* conditions, also known as their adaptive capacity (Smit and Pilifosova 2003). Second, the interaction of multiple risks and how adaptation to climatic stimuli occurs in light of these risks must be considered (Bryant et al. 2000; Hanemann 2000; Kandlikar and Risbey 2000; Pittock and Jones 2000; Smithers and Smit 1997a; Tol et al. 1998; Wheaton and McIver 1999). To address these challenges, some researchers document "actual" adaptations (Bradshaw et al. 2004; Chiotti et al. 1997; Smit et al. 1996; Smithers and Smit 1997b) and assess the vulnerability of agricultural systems. As noted in Chapter 4, such an approach attempts to understand the implications of climate change at a finer scale, from the perspective of the farm and farmer, and to generate information better suited to inform adaptation responses or policies (Burton et al. 2002; Smit and Pilifosova 2003).

This chapter presents and compares the findings of two studies of farmers' vulnerability to climate and other risks completed in British Columbia's Okanagan Valley between the fall of 2002 and the summer of 2004. There are many similarities in approach between these studies and the analysis of southwestern Ontario farmers described in Chapter 13. Reflecting the agricultural character of the Okanagan region, the studies presented here target apple and (wine) grape producers, respectively. The primary aim of the research was to identify: (1) the climatic and non-climatic risks that are pertinent to each sector, and (2) how farmers manage these risks.

After outlining the elements of the research project, including a description of agriculture in the Okanagan Valley, we provide results from the two studies. Findings point out the various climate and non-climate risks that producers identify as characterizing "bad years" and the farm-level adaptations employed to manage the associated risks. We conclude by discussing the meaning of these results in terms of producers' capacities to adapt to multiple risks and the nature of grape and apple growers' vulnerabilities.

Background for Research Project

The research presented here follows the steps for the "vulnerability approach" described by Ford and Smit (2004) and Lim and Spanger-Siegfried (2004) and outlined in Chapter 4. Both case studies were conducted in the Okanagan Valley in the southern interior of British Columbia, situated between two mountain ranges. Chapter 7 shows a map of this region (Figure 7.1) and provides a detailed summary of growing conditions. The combination of the rain shadow effects from the mountains and the moderating effects of the lakes that incise the valley creates the continental, semi-arid climate in the region that is ideal for fruit production. Although the valley receives little precipitation, particularly towards the south, agriculture is made possible by the use of low-cost irrigation, derived from snowmelt from the adjacent mountains (Neilsen et al. 2001).

The dominant agricultural activities are cattle production in the north and tree fruit production in the south and centre of the valley. Of all the horticultural commodities produced in the valley, apples are the highest grossing by a significant margin, earning over $73 million in 2002 (see Figure 12.1); this is six times the value of the next highest grossing tree fruit, cherries (BCMAFF 2002). Considering only wholesale values, grape production appears insignificant relative to apples, earning $739,000 in 2002. When the value after processing is calculated, however, wine grapes are the second highest grossing fruit in the Okanagan Valley (BCMAFF 2002). Further, the significance of the wine industry extends beyond the processed value, since the industry has a significant linkage to agri-tourism, which creates a number of spin-off effects for the valley's economy (Williams et al. 2001).

Data collection was conducted through a combination of semi-structured interviews and focus groups, using similar protocols. In both case studies, the apple industry and grape industry, participants were recruited through a combination of snowball and typical case sampling. The British Columbia Fruit Growers' Association provided an initial list of apple producers to contact, and these producers were subsequently asked to provide names of other potentially willing participants. The Pacific Agri-Food Research Centre

Figure 12.1

Value of fruit sales in the Okanagan Valley in 2002, with and without processing

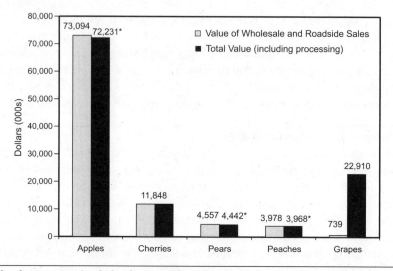

Note: In some cases (marked with an asterisk), packinghouse charges exceed the price paid by processors, resulting in lower total value than wholesale and roadside sales.
Source: BCMAFF (2002).

provided a short list of wine grape producers to contact. Other participants were identified either through snowball sampling or typical case sampling from a list of producers provided by the British Columbia Wine Institute (BCWI). The purpose was to obtain a sample of producers that was illustrative of the size and spatial distribution of operations in the valley. Tape-recorded interviews were conducted with a total of twenty-two grape producers and twenty apple producers. Focus group meetings for each commodity group were also held in three locations, Kelowna, Summerland, and Oliver. Participation rates in the focus groups ranged from three to five producers per meeting.

Producers responded to a structured series of questions regarding their experiences over the last ten years and prospects for the future, including their characterization of past *good* or *bad* years, and the farm management practices that they used to respond. The conditions identified in good years are considered opportunities and those identified in bad years are considered risks. In order to minimize bias in the responses, producers were asked about all possible conditions that affected them, and all management strategies employed. The interview guide did not prompt interviewees to discuss the climate variability or change until the very end, when producers were specifically asked about their views on climate change. The aim was to gain an understanding of producers' exposure to, and adaptive capacity for managing, multiple risks. It also provided a sense of where climate risks and climate change fit into producers' multi-risk decision-making environment.

Risk Exposures in the Apple and Grape Industries

Although both apple and grape producers work in the same region, with properties sometimes side by side, the risks that were of particular concern to each group differed (see Figure 12.2).

When asked about the conditions that would be classified as risks in the last ten years, more than three-quarters of the grape producers identified at least one weather condition. This indicates that climate risks feature prominently in farmers' decision-making environment, as weather is a manifestation of climate. The specific climatic stimuli that were identified as problematic were cool, wet growing seasons and extreme heat in the summer, which affect the vines' ability to mature the grape and achieve the desired quality in terms of sugar and acid concentrations. Grape production is also sensitive to frost and winter events because these can damage the vines and limit yield. While hot and dry seasons were typically described as ideal for grape growing, these conditions may also kindle forest fires in the region, a situation that can lead to smoke damage (which taints the grapes) in nearby vineyards and diminish the on-premise sales essential for many operations. Hot and humid conditions were also associated with

Figure 12.2

Risks that characterize bad years, as identified by apple and grape producers

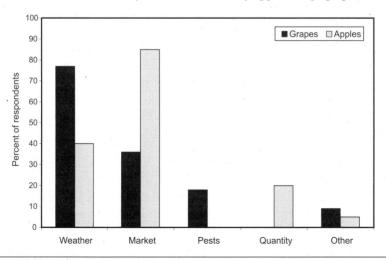

Source: After Belliveau et al. (2006).

outbreaks of mildew, and infected wine grapes cannot be used by or sold to wineries to make wine.

Just one-third of grape producers identified markets as a source of risk, which generally referred to their inability to sell their product, whether it was selling grapes to a winery or selling wine directly to consumers. Wine producers rely heavily on on-premise sales to winery tourists. In the past decade, tourism was reduced on a few occasions due to local and non-local events (e.g., the 2003 Kelowna forest fire, SARS (severe acute respiratory syndrome), and "9/11"). Reduced tourism is problematic for wineries, especially small wineries, because on-premise sales are more profitable than sales through distributors. Further, any loss in income negatively affects their ability to purchase grapes from independent growers, thereby creating a negative trickle-down effect for other grape growers.

Pests such as leafhopper and cutworm were also identified as risks, particularly by organic growers, who have a limited ability to respond to or control pest outbreaks. In general, it was noted that pests are not a major problem because the Canadian climate currently does not allow most pests to overwinter.

Like grape producers, apple producers identified market and weather conditions as their two primary concerns, although the relative importance of each differed. Nearly all the producers characterized bad years as ones with low market prices for apples, a factor that significantly affects their annual income. One apple grower stated: "The major factor is price. There are some years where Macs and Spartans were paying down to ten-twelve cents a pound on average, and Red Delicious down around six-seven cents a pound. You can't even get your production costs back with prices like that."

Many of the apple producers noted, however, that it is the combination of price and weather that characterized and distinguished good and bad years. Bad years were described as ones with low prices and weather conditions that either reduced quality or quantity. As in the grape industry, severe frosts are problematic for the apple industry because the tree may be damaged or killed and need to be replaced. Extreme heat is also a problem because it causes "sun scald," where the apples are essentially baked while still on the tree, and hence cannot be sold. Unlike grape producers, however, apple growers are primarily concerned with hailstorms because hail damages the appearance of the fruit. Whereas wine grapes are sent to the winery to be crushed, apples are primarily sold directly to consumers, and so any conditions (e.g., weather, pests, or diseases) that alter the appearance of the fruit make it unmarketable. While some of the spoiled apples can be salvaged and sent to the packinghouse for apple juice, this alone is not profitable for producers. Good, opportunistic years were the antithesis of bad years, characterized by high prices, high quality, and high yield, although producers did acknowledge the irony that producing a large crop can put downward pressure on commodity prices.

The disparity in the relative importance placed on market risks by apple and grape producers can largely be explained by the differences in the types of markets in which they sell. Apples are sold in a competitive and increasingly international commodity market, where prices continually fluctuate based on global supply and demand, and typically depreciate over the long term in inflation-adjusted terms. Hence, apple growers in the Okanagan Valley are "price takers," and the price they receive is set by the world market, not by individual producers. In the recent past, a key determinant of that price was apple production in neighbouring Washington state (i.e., the larger the crop produced in Washington, the lower the prices received by Okanagan growers). More recently, however, the relationship between regional production and price has been diluted by competition from other countries (Chile and New Zealand, for example) and by the entry of low-cost but high-quality apples from China. These new sources of competition were seen as most problematic for Okanagan producers, who foresee the increasing loss of their traditionally lucrative Asian export market.

Figure 12.3(a) shows the nominal prices for Macintosh and Gala apples between 1993 and 2002. The price for the traditional Macintosh variety has remained consistently low over this period relative to historical prices, fluctuating between 10 and 20 cents per pound. The Gala variety (which is relatively new for many producers in the Okanagan Valley) had substantial price fluctuations, ranging from 24 to 64 cents per pound during this period. While Gala prices have remained consistently higher than Macintosh prices, there is an evident downward trend (the grey line), as world supply has grown in the absence of an equivalent growth in demand. This downward trend is even more significant when prices are adjusted for inflation; in 2000, Gala prices were just 35 percent of those of the 1995 peak year.

Wine grape growers, on the other hand, have a different market structure. Okanagan wine is largely sold into a less competitive, specialty market where there is currently a high demand. Further, grape and wine producers who have both vineyards and wineries on their establishments are less concerned with grape prices than independent grape growers, because grape prices are equivalent to the costs of production. Assuming that the wine quality is decent, prices can be "set" to target certain market segments. Thus, grape and wine producers are "price setters." Even independent grape producers, growing grapes to be sold to wineries, experience less variable and depreciating prices than apple producers because independent grape producers are able to negotiate contracts with wineries prior to the growing season. In this case, prices are based on the quality of the fruit produced as well as the working relationship between the two parties, rather than a set world market price. Hence, not all growers receive the same price, but there is a degree of stability. Figure 12.3(b) illustrates this situation, using the example of prices for Chardonnay grapes between 1999 and 2003. The maximum and minimum prices indicate that on average this price has remained stable in the last five years.

Figure 12.3

Comparison of the prices of (a) apples and (b) wine grapes

a) Apples

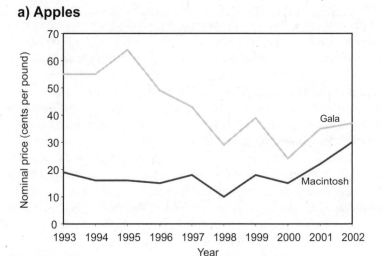

b) Grapes

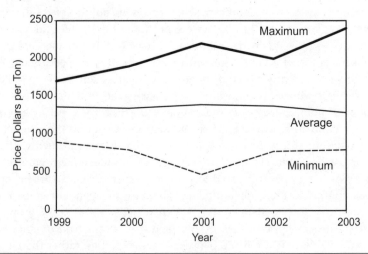

Source: BCMAFF (2004).

Risk Management Strategies Employed by Producers

In order to manage the climate risks that were identified, grape and apple producers employ a range of adaptations, primarily tactical, short-term farm production strategies (Table 12.1). In the event of a spring or fall frost, both

Table 12.1

Farm-level adaptations by grape and apple producers in bad years in the Okanagan Valley, British Columbia

Stimulus	Adaptations	
	Grape producers	Apple producers
Weather		
Cold, wet season	Drop crop, shoot thin Spray additionally for mildew Make sparkling wines Lower price of wine	
Frost	Irrigate Use wind machines Get crop insurance Choose an early-maturing variety	Irrigate Use wind machines Get crop insurance

Extreme heat	Irrigate	Irrigate
		Diversify household
		  income (spouse works
		  off farm)

Hail		Get crop insurance Send salvaged fruit to packinghouse
Fire/smoke damage	Get crop insurance	
Market		

Low prices		Tighten budget/reduce
		  spending
		Change crop varieties
		Produce high-quality
		  fruit
		Income stabilization
		  programs
		Diversify varieties
		Diversify household
		  income

Low tourism	Be more aggressive in other market channels Increase local sales	

groups use overhead irrigation to wet the grapes or tree roots so that as temperatures drop, there is a transfer of latent heat into the plants, protecting the berries and roots. Overhead irrigation is also used to cool down the vineyard or orchard in the summer, minimizing damage from extreme heat. Wind machines can also be used to reduce frost risk, especially when frost results from a temperature inversion; the machines pull down the warm air and raise the temperature in the vineyard or orchard. Wind machines are costly, however, and are economical only in larger operations.

In cold and wet seasons, the most common practice for grape producers is to drop a larger portion of the crop. A vine has the ability to mature a limited amount of fruit, an amount that varies by the type of grape and the availability of heat units. Hence, by dropping crop, more energy is available to the remaining fruit, allowing it not only to mature but also to achieve a higher quality. Dropping crop is a routine practice, performed not only to deal with poor weather conditions but also to obtain a quality product each season. The recurring motto of producers is, "We are going for quality, not quantity." One independent producer recalled dropping up to 60 percent of the crop in a poor season, while between 40 and 50 percent of the crop is dropped in most years. Wet seasons also create conditions conducive to mildew, so many producers increased the number of sulphur sprays in the vineyard. Wineries had the additional advantage of being able to compensate for poor-quality grapes by making different styles of wine or lowering the price of the product.

The majority of apple and grape producers (85 percent and 72 percent, respectively) buy crop insurance as a risk management strategy. Making a crop insurance claim was a common response to climatic conditions that reduced crop yield, either by spoiling the fruit or damaging the grapevine or tree (frost, hail, smoke damage). In the case of apple growers, crop insurance was the most viable option for responding to hail events; all of the respondents who cited hail as characterizing a bad year stated that they had crop insurance to cover the losses. Some apple producers even referred to hail insurance as "white harvester," because with it producers could receive the value of the lost crop as well as the value of the salvaged fruit used for apple juice, making their overall income, in some cases, higher than if they had had a good crop that year.

With variable market prices presenting a persistent risk in the apple industry, apple producers were able to identify a greater number of adaptations to manage market risks than were grape producers. The most common response for apple producers in low-income years is to reduce spending, tighten budgets, and not make unnecessary purchases, such as new equipment. Input costs are not cut, however, because the same amount of inputs, if not more, is required to produce a high-quality product. Similar to the grape industry, apple producers strive to produce high-quality products, because the higher the quality produced, measured by the size and colour of the apple, the higher the price received. An apple grower explained: "I don't normally hold back. I spend the money to produce a good crop even if it's a poor year because I am better off in the top end than anywhere else, regardless of what the market returns are."

Strategic, long-term adaptations that apple producers cited were to plant several varieties, under the assumption that not all varieties would do poorly in a given year, or to replant the orchard to higher-paying varieties. With the assistance of a government-sponsored replant program, growers in the valley were able to replace traditional varieties (such as Macintosh) with high-density, high-paying varieties (such as Gala or Jonagold). As Figure 12.3(a) shows, however, even "newer" varieties can suffer from declining prices, and this compels producers to continually replant in order to stay ahead of the price cycle.

Grape growers, on the other hand, had few responses for managing what market risks they faced. In times when tourism is down, winery operators tried, with limited success, to increase sales in other channels, such as through liquor stores and restaurants, as well as to local residents. Selling through the BC Liquor Distribution Branch is not a profitable option for small wineries, with some of their product selling at a loss, but the larger wineries, with greater economies of scale, sell through these other channels on a regular basis within and outside of British Columbia, and so are less affected by reductions in tourism. For the latter, reduced tourism is a lost opportunity rather than a significant risk.

Dynamic Vulnerabilities

The vulnerability of apple and grape producers is shaped by their exposure to climate and non-climate risks, and the strategies employed to manage risk, which is reflective of their broader capacity to adapt. Further, as the results of these two studies suggest, producers' exposure is largely characterized not only by the presence of multiple risks but also by the interaction of these stimuli. To illustrate, consider the two most commonly cited sources of risk, weather and market. Prior to the Canada/US Free Trade Agreement, BC grape producers grew French hybrid grape varieties, which are winter-hardy but produce a low-quality wine. The provincial government sponsored a "pullout" program in 1988, in which the French hybrid vines were replaced with the tender *Vitis vinifera* varieties (e.g., Chardonnay, Merlot, and Cabernet Sauvignon). This change enhanced the wine industry's international competitiveness, but simultaneously increased its susceptibility to winter injury. Similarly, the government sponsored a replant program in the apple industry, which prompted growers to replant old orchards with new, higher-paying, high-density varieties on dwarfed rootstock. Having dwarfed rootstock means that the roots are closer to the surface and are therefore more susceptible to frost and winter damage. The fruit is also more susceptible to sun scald, because the high-density varieties are less leafy, increasing the fruit's exposure to direct sunlight. Further, certain new varieties like Jonagold are inherently more prone to sun scald. Thus, in both cases, markets have been a fundamental driver of adaptation, which in turn has increased vulnerability to certain climatic stresses.

The interaction of market and climate risks also influences the way in which producers are vulnerable. Wine grape producers repeatedly cited cold, wet summers as conditions that led to a bad year. These conditions do not seriously affect the quantity of grape production, but they reduce the quality of the fruit, and hence of the wine. This, in turn, reduces a winery's price premium or may even wipe out sales. Thus, in this case, producers are not vulnerable to the climatic stress itself. Rather, it is the expression of the climatic stress as a market risk that characterizes the exposure. Apple growers are similarly concerned with quality, in terms of the cosmetic appearance of the fruit and its size and colour; thus, they are sensitive to the conditions that affect quality, such as hail and extreme heat. Since apple growers are price takers and are not able to negotiate prices, they are more sensitive to reductions in quality because as quality goes down, prices drop considerably; a loss of even one size in an apple represents a significant price difference.

The interactive effects of risks are not limited to climate and market. For example, government policies in place in the wine industry are not inherently risky, but they affect producers' exposure to weather and market risks. This is most evident in the "markup-free delivery" policy, which allows producers to capture the 100 percent markup on their product through on-premise sales, whereas this markup is captured by the government in Liquor Distribution Board sales, leaving little profit for the wineries. The existence of this policy means that wineries, particularly small ones, rely on these on-premise sales in order to retain enough profit to stay in business. This reliance in turn means that producers are vulnerable to reductions in tourist traffic.

The availability of government support programs generally enhances adaptive capacity by reducing producers' vulnerability to risks. Apple producers, however, identified a drawback to the new Canadian Agriculture Income Stabilization (CAIS) program. Producers who diversify their varieties or crops to reduce their sensitivity to weather or price risks may, ironically, be worse off in the event of a price downturn in, or weather-related damage to, one variety or crop, in that their resulting overall decline in income may not qualify them for assistance under CAIS. Thus, by strategically reducing their exposure to risk, producers are potentially reducing their overall adaptive capacity. The same is true for operations that diversify their location, such as wineries that purchase vineyards in two or more locations. Unless the vineyards are covered under separate crop insurance claims, a loss of crop in one vineyard may not be sufficient to support a claim and producers must bear the loss.

Apart from this shortcoming, the availability of government support enhances the adaptive capacity of producers with regard to weather-related conditions that affect yield, such as frost, winter injury, smoke damage, and hail. For weather risks that are less serious and that affect quality without reducing yield (e.g., extreme heat and cold, wet seasons), there is no safety net, however. Farmers can try to minimize the negative effects through tactical farm production strategies, such as dropping crop and irrigating, but in the end, if quality is reduced and the product is deemed less valuable in markets, producers have a limited capacity to respond. Again, faced with reduced market prices, apple producers can do little more, in the short term, than tighten their budgets and try to draw on non-farm sources of income. Although government income stabilization programs exist, they tend to be available only in extreme income disaster situations.

From this discussion of exposure and adaptive capacity, a number of insights can be gained regarding the nature of grape and apple producers' current vulnerability to a variety of risks. Both grape and apple producers are highly sensitive to severe winters and spring and fall frost because these can injure the vines/trees and reduce cropload. In addition, grape producers are sensitive to smoke damage and apple growers to hail events because both conditions spoil the crop for the season. For these risks, government support programs are available to enable producers to cope with and to recover from the loss. For the relatively modest but still troublesome climatic stimuli that reduce the quality of the fruit (e.g., cool and wet conditions), the primary adaptations include tactical farm management practices that attempt to minimize such negative effects; should these adaptations prove ineffective, however, and the product suffers a heavy price penalty in quality-conscious markets, apple and grape producers have little recourse. Thus, ironically, the climatic stimuli that are most difficult for producers may not be those that are extreme and most physically damaging but rather those that produce no change in fruit quantity but reduce quality to a point where markets notice. Further, the specific markets within which producers sell their product influence both the risks to which they are exposed and their ability to adapt. Producers who sell into highly competitive commodity markets are not only more exposed to variable and typically depreciating prices but also more likely to be impacted by this quality-induced price penalty. While wine grape producers, and especially independent growers, are not immune to such circumstances, it is primarily Okanagan apple producers that suffer such a fate.

Future Vulnerability

The above discussion has outlined how producers are vulnerable to current climatic variability in the context of other sources of risk. To gain insight into how producers might be vulnerable to a future climate change, inferences can be drawn from future climate scenarios or probabilities of change in the exposures that were identified by producers. Should average temperatures increase by 1.5-4.0°C by 2050 in the Okanagan, as climate scenarios suggest, there may be opportunities for grape producers in growing later-maturing varieties and achieving higher quality on a more consistent basis as a result of the additional heat units available. Apple producers may also be able to replant to new varieties and produce larger fruit. Warmer temperatures may also reduce the risk of frost and winter damage for both commodities, particularly in the fall. If spring temperatures increase, budbreak for both grapes and apples might occur earlier, and hence the risk of spring frost may be the same. An earlier budbreak would mean that both apples and grapes have an opportunity to mature before the fall frost period. On the other hand, higher temperatures may increase the occurrence of extreme heat events, and hence sun scald.

A change in climate may also affect producers' adaptive capacity, making certain adaptations ineffective or simply unavailable. This is especially so in the case of irrigation, which may be compromised in the future through reduced water availability. It is expected that climate change will affect the hydrology in the valley, changing the form and timing of precipitation throughout the year (Merritt and Alila 2004) and limiting producers' access to water for irrigation at critical times throughout the season (see Chapter 7). This problem will be exacerbated by increased competition with residential users, whose population is projected to increase by 63 percent between 2001 and 2050 (Neilsen et al. 2004a). Since agriculture in the valley is reliant on irrigation, and because irrigation is an adaptive strategy to manage several weather risks, a reduction in the water supply represents a key vulnerability for both apple and grape producers.

Conclusions

In this chapter, we have presented and compared the findings of two separate studies of Okanagan Valley apple and grape producers' vulnerability to climate and other risks. In keeping with the first two steps of the vulnerability approach, data have been presented to identify: (1) the climate and non-climate risks that are pertinent to each sector, and (2) how farmers seek to manage these risks through various adaptations. These data provide insight into how producers are currently vulnerable to climatic variability and how this vulnerability can change over time as the risk exposures, including climate, change and as producers adapt to these risks.

The first conclusion is directly related to observations noted in several earlier chapters in this volume, namely, that producers experience and adaptively respond to a number of interrelated risks (and opportunities). Climate risks, for example, tend to be expressed through, or in relation to, market risks, as was especially evident in the case of the apple producers, for whom weather-induced declines in quality resulted in significant losses in prices received from highly competitive and increasingly international apple markets. Further, the interaction of climate and market risks illustrates the dynamic nature of vulnerability, with apple and grape producers tending to respond to market demands by replanting with varieties, which in effect alters the degree to which they are susceptible to extreme climate events such as a winter deep freeze.

A second conclusion from the work stems from the recognition of the differing adaptive capacities among commodity groups, in this case apple and grape producers, and among individual producers. Grape producers have more options for responding to, or at least coping with, weather-induced quality declines, given their ability to drop crop, produce sparkling wines, and augment income through agri-tourism, compared with apple producers, for whom quality declines were generally hard to mitigate. Adaptations exist for both apple and grape producers to avoid quality declines brought about by frost or extreme heat, but not all apple or grape producers have wind machines or overhead irrigation; indeed, some adaptations are limited to just the larger and more capital-intensive operations. In any case, the ability to adapt may be compromised in the future with changes in hydrology affecting the availability of water for irrigation in the Okanagan Valley.

If weather events are sufficiently extreme that entire crops are lost or whole trees or vines are destroyed, all producers, whether big or small, tend to have ample adaptive capacity in the form of crop or tree/vine insurance. This finding points to a final and somewhat counter-intuitive conclusion of the research. For sectors like apples and grapes, where markets heavily penalize output of just slightly inferior quality, producers are less vulnerable to extreme weather events than they are to non-extreme but troublesome weather, such as prolonged cool and wet conditions, which typically reduce output quality. This is particularly problematic for producers who sell into commodity markets, such as Okanagan apple producers, where prices are set by a world market rather than producers. In the future, the vulnerability of producers to climate change and variability, particularly in the apple industry, will be influenced not only by climate changes but also by the continued internationalization of apple markets and the risk management strategies employed to cope with the changes.

Acknowledgments
The authors gratefully acknowledge financial support from the Government of Canada's Climate Change Impacts and Adaptation Program, the Royal Canadian Geographical Society, the Social Sciences and Humanities Research Council, and the University of Guelph. Special thanks to Bronwyn Sawyer for her contributions to the research project. For his technical support and skills, the authors would also like to thank Luke Powell. Further, the authors extend sincere thanks to the many apple and grape producers of the Okanagan Valley who gave of their time for interviews and focus groups, as well as to Drs. Denise Neilsen and Pat Bowen at Agriculture and Agri-Food Canada's Pacific Agri-Food Research Centre and to the leadership at the British Columbia Fruit Growers' Association and the British Columbia Wine Institute for making the field work possible.

13

Vulnerability and Adaptation to Climate Risks in Southwestern Ontario Farming Systems

Susanna Reid, Suzanne Belliveau, Barry Smit, and Wayne Caldwell

Both the context- and process-based approaches contain references that imply a need for understanding farmers' perceptions of climate and climate change, how such perceptions are affected by climatic conditions, the adaptive strategies that are available to farmers, and the constraints and opportunities for enhancing farmers' adaptive capacity. Indeed, as noted in Chapter 4, such information is a basic first step in examining sector vulnerability and adaptive capacity. The processes of adaptation involve several groups of stakeholders. The scientific community can contribute research and information; governments can facilitate adaptation and identify and eliminate barriers to adaptation; and farmers can become informed about risks and opportunities of a changing climate and adapt their practices to moderate hazards and to realize opportunities (Dolan et al. 2001; Smith et al. 1998; Smit and Skinner 2002).

This chapter focuses on exposures or risks at the farm level and the management of risks associated with climatic variations and changes. It documents climate-related conditions experienced by farmers and their farm-level responses to risks and opportunities for their operations. We provide an example of the process-based approach to assessing agricultural producers' ability to adapt to climate change, drawing from a study of farmers in southwestern Ontario that was conducted during 2003 (Reid 2003).

The sections in this chapter provide the conceptual framework for the study, background information on the case region, the methods employed, a summary of the results, and a discussion of implications for understanding capacity in the particular group studied.

Conceptual Framework

Based on the elements of the vulnerability approach (see Figure 4.3) and on conceptualizations of farmer decision making (Smit et al. 1996), the conceptual framework in Figure 13.1 was developed. As the framework suggests, individual farms (and farmers) are exposed to variable climatic

Figure 13.1

Model of forces influencing farm vulnerability and adaptation strategies

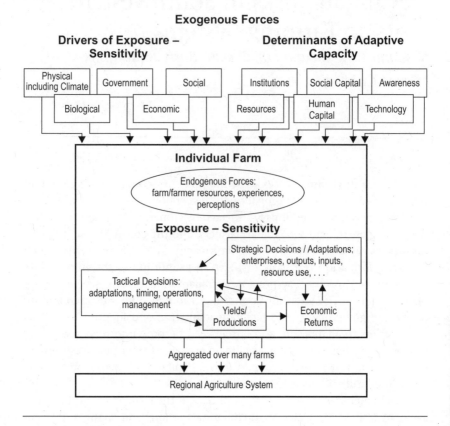

conditions at the same time that they are exposed to a variety of exogenous non-climatic conditions and stimuli. These conditions combine to influence production and other facets of farming operations, and they also influence adaptive capacity or the ability of the farm (or farmer) to respond to changing conditions, including those related to climate.

These external forces (biophysical, economic, social, etc.) are experienced through the particular (endogenous) characteristics of the individual farm and farmer. Farmers are differentially exposed and sensitive to external forces according to the particular characteristics of the farm – its location and land type; farm enterprises; and scale (financial, among others). Decisions are made in light of these forces over both the long-term (strategic) and short-term (tactical) time horizons. These decisions may include measures or strategies that represent adaptations to climate risks.

Figure 13.2

Location of Perth County, Ontario

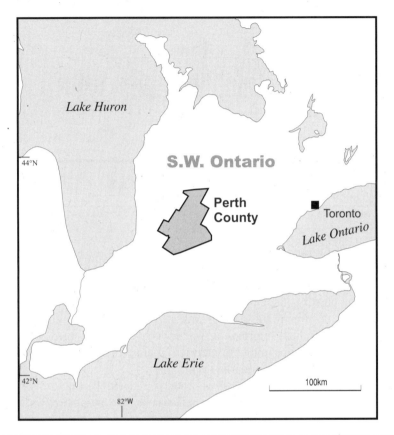

The case study described here was conducted in Perth County (Figure 13.2), situated in southwestern Ontario, a distinct agricultural region that contains over half of the Class 1 agricultural land and generates almost one-quarter of all farm revenue in Canada (Government of Ontario 2004). Perth County is very suitable for agriculture, with approximately 13 percent of its employment in that sector (Cummings and Associates 2000). Production includes cash crops (soybeans, corn, and wheat) and livestock.

The climate regime is characterized by warm, humid summers and cold, snowy winters. Temperature and moisture vary considerably both within seasons and between years, however. For example, moisture variability for the reproductive period[1] is relatively pronounced and could reflect a general trend (Figure 13.3).

Figure 13.3

Trends and variations in moisture balance for the reproductive period, London Station, 1895-1997

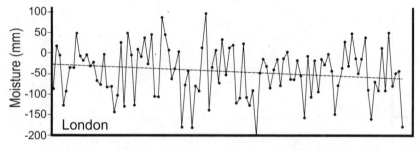

Moisture balance for the reproductive period

Source: After Reid et al. (2007).

The Perth County study was designed to identify those attributes of climate and other conditions to which farmers are vulnerable; to characterize the nature of those vulnerabilities; to assess existing adaptive strategies; and to consider possible adaptation options for dealing with future climate risks. The data to document those items reside in the experiences and perceptions of farmers. Such information was collected primarily through interviews and focus groups to identify farmers' management decisions and the forces and processes that underlie them.

Twenty-five in-depth interviews were completed with Perth County farmers at their farms. Interviewees were selected using a purposive, snowball sampling approach to represent the main types of farms in the county. With a focus on the last ten years, farmers were asked to identify and assess "risks and opportunities facing agriculture." Rather than lead with climate or weather, questions were open-ended, inviting specification of all types of exposures and sensitivities. Only at the end of the interview were respondents asked about climate change.

In addition, four focus groups were held to provide a cross-check on the findings from the interviews. The sessions included discussion of the information gathered from the interviews to see whether there was general agreement and whether any further insights could be added. The same criteria for selecting interview participants were used for the focus groups to gain representation of the main farm types, and the same open-ended approach to the discussion was also employed.

These data-gathering techniques enabled the researcher to gain insights into how Perth County farmers make decisions and respond to climate and growing season conditions. Such methods are well established as a means of understanding decision making in light of changing conditions (Flick 1998; Palys 1997; Smit and Wandel 2006), and they have been employed in other studies of climate change and agriculture in Canada (Belliveau et al. 2006; Brklacich et al. 1997b; Chiotti et al. 1997).

Summary of Results

In keeping with the discussion in Chapter 4, the findings have been organized according to (1) the conditions that represent current vulnerability, including both exposure-sensitivities and adaptive strategies, and (2) future vulnerability, including future exposures and adaptive capacity. The results are presented under two subtopics, namely, current and future vulnerability. Results from the study described here provide substantial detail for current vulnerability but offer only more general views on the future.

Current Exposure

Information collected from farmers describes the complex socio-political environment within which farmers make management decisions. The forces and events that influence farmers and their decisions do not occur separately or independently – every year, farmers are influenced by a combination of many conditions.

As noted in Chapter 11, a multitude of forces internal to the farm family and business prompt certain choices and constrain others. Based on information from Perth County farmers, decisions of family members to join or leave the farm, accidents, and financial circumstances all influence the vulnerability of the farm, and many prompt adaptations. Long-term and short-term management decisions are moderated by a number of factors. For instance, farmers' management style (such as early or late innovators) and interest in expanding their operations greatly influence management decisions with respect to external pressures.

Table 13.1 lists the exogenous forces influencing Perth County farm operations. With no prompting, participants in the interviews and focus groups frequently mentioned weather as the first exogenous influence, noting that good weather (heat, sun, and moisture at "the right time") significantly increases crop yields, and hence economic returns. As well, they indicated that wet weather and drought reduce crop yields; wet springs and late springs make planting difficult; cool summers (low heat units) reduce crop yields; and wet falls make harvesting difficult.

A good year is a combination of a lot of things ... I think we did a pretty decent job last year, and growing conditions were excellent, could do the same thing this year and have 30% less crop if we don't get heat and the

Table 13.1

Exogenous forces influencing farmers in Perth County, Ontario

Exogenous Force	Example
Climate	Heat, moisture, timeliness of heat and moisture, absence of extremes
Soil, pests	Soil moisture, aphids
Economic/market conditions	Product prices, input costs, market timing
Government policies	Legislation, regulations, safety nets, income support
International trade, foreign government policies	Foreign subsidies, world markets, trade agreements

right growing conditions ... If you don't get the right kind of weather when corn is tassling it is going to reduce your yield. All these factors that need to happen in a timely manner. You can't have a month of really cloudy weather conditions, and then a hot summer ... You have to have good, favourable weather conditions. (Hog farmer)

Crop pests and disease and the health and productivity of animals are also factors contributing to the quality of a year. Farmers described a strong connection between crop pests and climatic conditions. Several farmers noted that soybean aphids, which devastated many soybean crops, first appeared in the drought of 2001. Farmers also identified a connection between warm winters and pest infestations.

We had aphids two years ago ... it was a dry, dry year. And the explanation I got, south winds came up from the states more than they usually do because of the wind direction ... Got less than half expected yield. Usually get over the 40, 50 bushel (per acre) and that was down around 20. (Dairy and cash crop farmer)

Farmers experience climate and other environmental conditions in the context of other forces. Agriculture is as much an economic activity as it is a biological one, and Perth County farmers noted their sensitivity to market and other economic conditions.

The market is the next big thing after weather. We depend on what those buyers want to pay for our cattle. A few cents on a hundred weight can make a big difference. I tend to keep my cattle spread out so I am shipping on different markets ... sure can make an extra one, two, or three thousand dollars when the market is better. (Beef farmer)

Government policies at various levels influence agricultural practices and the economics of farming. Perth County farmers are sensitive to these political conditions (such as environmental regulations), particularly as they affect operations and income.

We are hoping that with the nutrient management legislation we can get assistance. Cure us or kill us. If there is no funding, and they want us to do something, we won't be able to keep doing it. We want to do it, but we have to see our way fit to do it. (Dairy farmer)

The agricultural sector includes safety net programs, managed through governments, to provide various types of insurance or income stabilization support when farmers experience adverse conditions.

We have taken advantage of GRIP [Gross Revenue Insurance Plan], NISA [Net Income Stabilization Account]. National tri-partite years ago when I was in pigs. GRIP I think is gone now. We have depended on NISA quite heavily. Last year we had a bad cash flow year, just the way things worked out ... a lot of things we did affected our cash flow. It's a good way to invest money. (Beef farmer)

Foreign government programs and international trade arrangements also represent exogenous forces to which farmers find themselves exposed. Trade regimes and global supply and demand conditions influence prices for products. Likewise, state support for agricultural activity in foreign countries affects market access and prices.

We have had some assistance from the government, but for every dollar we have got, the Americans have got $2 and the Europeans have got $3. This past year, the US Farm Bill increased that. Our government has made more moves to assist us. Other governments which assist their farmers, keeping the world price of the commodity low, hand out more money to the farmer that our government can't match. Our government has been faced with a choice. And there has been some response to it over the last year, from that if they want to keep the crop farmers alive they are going to have to find some money for it. (Poultry and cash crop farmer)

These results show that climate and weather conditions are very important to Perth County farmers but so too are the environmental, economic, and political forces.

Current Adaptations

Farmers in this study of Perth County recognize that certain climatic conditions are directly and indirectly detrimental to their business and seek to deal with those exposures in their ongoing strategic and tactical management decisions. Table 13.2 provides examples from Perth County farmers of specific types of climatic exposures (weather), associated farm sensitivities, and farmers' adaptations. The sensitivities and adaptations vary according to farm operation (crops, livestock type, physical conditions).

Table 13.2

Examples of climatic exposures, sensitivities, and farm-level adaptations in Perth County, Ontario

Climatic exposures	Sensitivities	Farm-level adaptations
Good weather (hot, sunny weather with timely rains)	Crop yields, feed costs (livestock)	Stockpile hay Improvements in equipment Pay down debt, buy more land
Hot summers	High heat units, crop maturity	Opportunity for longer-day corn, beans Higher yields
	Beef cattle weight gain slows	Ensure shade available Use barn if it is cooler there
	Too hot in chicken barn, quality, losses	Increase ventilation Upgrade ventilation
	Heat reduces hog conception rate, semen quality reduced	Overbreed More diligent pregnancy checks More artificial insemination Air conditioner for barn
Drought	Moisture deficit, crop yields, aphids on soybeans, lost yield	Reduce inputs Crop insurance In future, would spray
	Weeds don't respond to herbicide	Roundup is safer in drought years Use different sprays or none at all
	Hay doesn't grow back, reduced hay harvest	Only get first and second cut Don't sell hay Buy grain
Climatic exposures	Sensitivities	Farm-level adaptations

▶

◄ *Table 13.2*

Wet	Trouble making hay	Pasture cattle on hay field Delay or don't cut Kept hay from previous year Tile-drained field
	Rust in grain, quality	No response
Wet fall	Lost bean crop	Used for feed Stopped growing beans Tiled fields
	Poor corn crop	Crop insurance claim Ploughed crop under Increased storage capacity to stock- pile feed from good years Made cob meal, silage
Heavy rains, flooding fields	Excess moisture, erosion, saturation	Pasture management practices Tile fields; repair tiles No-till, protect soil Take hollows in fields out of production (plant grass)
	Standing water in field, crop rotted	No response
	Flattened soybean field	Put in a berm to slow water, reduce erosion
	Low heat units	Plant several maturing varieties of corn, spread out risk
	Poor-quality crops	Crop insurance Bought feed
Cold, wet weather	Corn rust, corn wouldn't dry down	Insured corn the following year Planted lower heat units the following year Used corn for feed
	Trouble making hay	Neighbours gave them hay Sold land the following year
	Scouring in beef calves	No response
	Hog feed consumption is higher	Increase feed
	Lost winter wheat	Crop insurance claim
Colder winter, no snow	Possibility of insect problems	Watch for insects (precaution)
Warm winters	Risk of alfalfa being damaged	No response

Farmers respond through land management and tillage, crop choice, feed storage and purchase, and crop insurance.

Years with favourable climatic conditions produce benefits of yield and income and can encourage adaptations such as stockpiling feed, investing in equipment and land, and paying down debt. While high levels of heat are generally beneficial (often providing opportunities for higher-yielding crop varieties), very hot summers are problematic for livestock (cattle, poultry, and hog) production. Extreme heat requires tactical measures such as shade provision, increased ventilation, and modifications to hog breeding.

Drought conditions and the associated pest incursions have prompted adaptations ranging from altering spray applications through purchasing crop insurance. Excess moisture diminishes yields and has led to tactical adaptations such as altering hay cutting and use, converting cash crops to feed, and ploughing crops under. Longer-term strategies relate to the use of crop insurance, increases in storage capacity, modification of tillage practices, and installing tile drainage in fields.

Farmers in Perth County have developed a wide variety of farm-level management practices to adapt to weather-related effects and risks. This range of responses indicates that farmers have considerable adaptive capacity and are already adapting to climate risks. There are still questions about whether the current adaptive strategies will be effective as climatic conditions change (including alterations in the frequency and severity of extremes).

Future Vulnerability

Based on data gathered, farmers appear to be very aware of the immediate risks to their operations from climate- and non-climate–related forces but are generally unaware of and/or unconcerned about the influence of long-term climate change. This finding is consistent with other reviews of farmer perceptions of climate change (Bryant et al. 2004; Smit et al. 2000a). One of the reasons for this indifference is that "climate change" may be thought of as very long term (several decades beyond the time horizon for their management decisions) and relating mostly to gradual temperature increases ("global warming"), with little reference to changes in the frequency and severity of heat and moisture extremes.

Climate science has now demonstrated that several weather/climatic conditions that are deviations from the "norm" and that affect Perth County farmers have altered in frequency and severity with climate change. Some of these may be welcomed. Increases in the frequency of high-heat growing seasons and decreases in excess spring and fall moisture could present opportunities for farmers. On the other hand, increasingly frequent and widespread drought conditions, periods of extreme heat, lack of protective winter snow, milder winters beneficial to pests and diseases, and more frequent intense rain storms all represent problematic exposures for many farm operators.

The lack of interest regarding climate change does not necessarily mean that farmers are highly vulnerable to future climate risks. In response to inter-annual climatic variability, farmers have developed a range of tactical and strategic, anticipatory, and reactive adaptations to deal with their vulnerability to climate. Government programs offering financial assistance also provide support following weather-related disasters. Experience in adapting to climatic variability offers a certain level of preparedness for the expected effects of long-term climate change on agriculture.

A review of Perth County farmers' adaptive capacity identifies a number of factors (sometimes called determinants of adaptive capacity) that affect the adoption of adaptation measures and strategies (Table 13.3).

Table 13.3

Factors that affect farmers' adoption of climate change adaptive strategies

Determinants of adaptive capacity	Influences on adaptation	Barriers relating to adaptation
Awareness	Farmers recognize weather as a very important condition Type of farm affects risks (e.g., crop farmers are more concerned about climate change than livestock operators) Internet, seed companies important information sources	Farmers are unaware/ unconcerned about climate change Limited extension services on climate change effects and adaptation Farmers do not link climate change to their immediate concerns Farmers are confident in their own abilities and generally accept their own limitations in the face of extreme weather conditions
Technology	Equipment increases efficiency of production in good weather New seed varieties more resilient to pests No-till improves soil conditioning	Decisions require research, time, experience High cost of new technology Technology can increase output in good conditions, but not help in extremes (e.g., high-yield varieties in drought) More specialization reduces resilience from diversification

▶

◄ *Table 13.3*

Resources	Financial viability facilitates adaptation investments Access to capital and sustained low interest rates permit management expenditures Owning equipment allows for effective timing of field work	Ontario net farm income declining Farmers retiring soon not likely to invest in changes Small farms have less access to capital for new technology Volatility of hog market puts hog farmers at financial risk Farmers who rely on custom operators have less control of timing of field work
Institutions	Some regulations indirectly encourage adaptation	Governments do not yet perceive climate change as a risk to producers Lack of communication from governments regarding climate change risks and adaptation options Some regulations indirectly restrict adaptation Differential subsidies affect Canadian farmers' incomes
Human capital	Farming community is skilled, innovative, experienced, and knowledgeable Early innovators try out new practices and technologies Young farmers bring new ideas	Late innovators wait until technology is proven New farm generation not working in the farm business can limit growth and incorporation of new practices Staying small restricts incorporation of new technologies
Social capital	An established agricultural community, social networks, and organizations Established agricultural industry, infrastructure, and supply system	Farms are increasing in size and decreasing in number Young people are not entering the sector
Risk spreading	Government safety net programs Farmer diversification Feed poor crops to livestock Off-farm income	Safety net programs are being revised Modernization of agriculture has encouraged specialization; technology and marketing can discourage diversification

Awareness of farm-level risks and opportunities related to climate change is a key ingredient in adaptation. If farmers hold the view that climate change is essentially irrelevant to them, adaptation measures are not likely to be adopted. Government agencies are important as sources of information and as providers of safety net programs. Current risk management instruments such as crop insurance are likely to be affected by the number of claims associated with extreme weather events, especially if they increase in number and severity.

Technology offers some prospects for adaptations as long as financial resources permit. As described in detail in Chapter 8, the forces of marketing, technology, and finance have tended to push for fewer and larger farms with more specialized enterprises. Many Perth County farmers have shown that diversification is an effective strategy for dealing with unpredictable risks, particularly those related to climate/weather. Social and human capital are important aspects of an agricultural community's capacity to deal with variable conditions (Wall and Marzall 2006), but as those communities continue to experience decreases in the numbers of young people, there may be diminished human and social resources for adaptation.

Conclusion

The process-based approach described in this chapter provides insights into agriculture's sensitivity to climate change, including variability and extremes. The research has identified the attributes of climate (and of climate change) that are problematic for farmers in the study region. According to the participants, several factors are as important as growing season temperature and precipitation; also key are the frequency and duration of droughts, the timeliness of moisture with adequate heat, the length of the growing season, and adequate snow cover. The research also shows that farm operations experience climatic stimuli together with numerous non-climate forces, and that these are moderated by the particular circumstances of the farm and farm family. Management decisions, including risk management decisions, are made on an ongoing basis in light of many forces and conditions. It is unlikely that risks associated with climate change will be managed independently of ongoing management practices.

Farmers in Perth County currently employ a wide range of adaptive practices to deal with climate and weather risks, and many are considering adaptations to deal with longer-term climate change. While they appear to have considerable adaptive capacity, some of the projected changes in climate (such as increased frequency or severity of droughts) will likely require additional adaptive capacity. Such an enhancement may demand action on the part of public agencies (through awareness raising, policy review, technology development, etc.) and individual farm operators.

Acknowledgments

This research was sponsored by the Social Sciences and Humanities Research Council of Canada, the Ontario Ministry of Agriculture, Food and Rural Affairs, the Climate Change Impacts and Adaptation Program of the Government of Canada, and the Canada Research Chair Program. We are grateful to the many Perth County farmers who participated in the study.

Notes

1 The reproductive period refers to the time between the start of the growing season and the date that 70% of the CHUs have been accumulated; it represents the stage in plant growth when crops are more sensitive to available moisture.

14
Community-Based Watershed Management as an Agricultural Adaptation to Climatic Extremes in the Canadian Prairies
R. Cynthia Neudoerffer and David Waltner-Toews

Chapter 4 explains how and why the notions of vulnerability and adaptive capacity have recently emerged as central to the discourse on adaptation to climate change. As noted, both concepts are not new per se but have a long history in the literature on natural hazards and disasters. Building social and ecological resilience has recently been proposed as one possible lens through which to view the process of decreasing vulnerability and increasing climate change adaptive capacity. This chapter provides preliminary observations from a case study where community resilience for dealing with climate change impacts is featured.

The concept of resilience is also not new to the field of climate change vulnerability and adaptation, but it has rarely been explored beyond a general idea or concept. In contrast, within the fields of ecology and natural resources management, a rich development of the concept of resilience, especially the notions of social and ecological resilience, has taken place over the past thirty years. In Chapter 9, Henry Venema uses elements of resilience to argue that agricultural policy needs to consider fully possible outcomes for climate change adaptation and sustainability.

Resilience is commonly defined to mean the ability to bounce back from a shock or perturbation. This definition, as Holling has extensively illustrated, is appropriate for simple systems (linear, predictable) organized around a stable equilibrium (Holling 1986, 2001). The types of linked social and ecological systems that comprise the basic units of agricultural adaptation to climate change are anything but simple, however. Indeed, agricultural systems are "complex" in the sense that Holling describes – non-linear; dynamic with multiple possible equilibrium states; characterized by feedback loops, uncertainty, emergence, and surprise; and inherently unpredictable. For such complex systems, resilience is defined as the magnitude of shocks a system can absorb and still remain within a given state; the degree to which the system is capable of self-organization; and the degree to which the system can innovate, experiment, and learn (or build adaptive capacity) (Carpenter et al. 2001).

It has been observed throughout this volume that climate change adaptation research needs to consider multiple spatial and temporal scales as well as the socio-economic and environmental conditions that are inherent to any understanding of adaptive capacity/vulnerability. Given the wide range of different perspectives that must be brought to bear on the question of adaptation to climate change in Canadian agricultural systems – producers, agribusiness, environmental groups, and governments, to name a few – we believe that it is useful for researchers, policy makers, and other decision makers to have a number of different tools (or perspectives or theoretical lenses) at hand through which to view the issues. Our hope is that a social and ecological resilience perspective may serve to complement and strengthen the vulnerability approach to climate change adaptation in Canadian agricultural systems in general.

Given that building social and ecological resilience is one way to reduce vulnerability and increase adaptive capacity in the face of climate change, there is merit in carefully examining resilience and seeing how it develops in communities. In short, understanding the past can help with preparations for future adaptation to climate change. The chapter begins with background information about the research and includes a discussion of the relevant theoretical concepts. Details from the case study are provided, demonstrating how social and ecological resilience can emerge in the face of extreme precipitation events in a rural community. The chapter ends with a discussion of findings and implications for employing a vulnerability approach featuring resilience.

Background of the Research Project

The research presented here follows the elements of Figure 4.3, namely, the need to appreciate the historical and/or recent exposures and adaptations as a basis for determining future adaptive capacity. As such, the research presented in this chapter is motivated by three key questions:

- How are rural communities in the Canadian Prairies currently organizing to adapt to soil and water problems caused by too much (spring flooding and summer storms) and/or too little (drought) precipitation?
- What conservation actions are farmers taking (or have farmers been taking)?
- How are they organizing to address local concerns?

From a climate change perspective, projections for the Canadian Prairies generally suggest that warmer temperatures, drier conditions, greater incidence and severity of both drought and extreme precipitation events, and reductions in water availability and quality will prevail in future years (Chiotti 1998). As pointed out in Chapter 6, such projections for the Prairies need to be understood against the historical climate of the Prairies. Drought is already a significant and distinguishing climatic feature of the Prairies.

What this means is that our current understanding of "normal" climate on the Prairies may actually be a "wet" cycle, and this is regardless of climate change. There is, therefore, the potential for a "double impact" effect for the Prairies – a return to more "normal" drier conditions *and* the imposition of yet drier and more variable conditions induced by climate change.

Theoretical Concepts

The research uses two main theoretical concepts: vulnerability (V) and resilience (R).

$V = f$ {exposure; adaptive capacity}

$R = f$ {ability to absorb change; self-organization; adaptive capacity}

Vulnerability is defined as a function of both exposure to a climatic stimulus – in our case an extreme precipitation event – and the adaptive capacity to manage or reduce the impact of that event. Vulnerability is a location-specific measure that is not static; rather, it changes over time, sometimes increasing, sometimes decreasing, depending on the interaction of its two constituent parts (Smit and Pilifosova 2003).

Resilience is defined as a combination of: (1) the magnitude of shocks that a social-ecological system can absorb and remain within a given state, (2) the degree to which the social-ecological system is capable of self-organization, and (3) the degree to which the social-ecological system can learn, experiment, and innovate or build adaptive capacity (Carpenter et al. 2001; Folke et al. 2002). Thus, like definitions of vulnerability that encompass both biophysical and (socially constructed) social vulnerability, resilience is understood in this research to incorporate both biophysical changes in the landscape and the social institutions required to manifest those changes.

The idea of resilience is captured in Holling's "adaptive cycle or Figure 8" diagram, as shown in Figure 14.1. This depiction emphasizes the cyclical nature over time through the four phases of exploitation, conservation, release, and reorganization (Gunderson and Holling 2002). Holling's adaptive cycle or "Figure 8" metaphor plays a key role in organizing the concept of ecological resilience (Gunderson and Holling 2002). In early work on ecological resilience, Holling (1986) noted that dynamical ecological systems do not tend towards one stable or equilibrium position, but rather have the possibility of flipping into a number of (multiple) stable states.

Figure 14.1

Holling's adaptive cycle

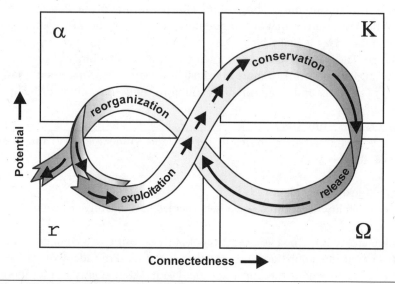

Source: After Gunderson and Holling (2002).

The likelihood that an ecological system will remain within a desired state is related to slowly changing variables that determine the boundaries beyond which disturbances may push the system into another state. Holling also argued that within any one system state, a dynamic ecosystem would pass through the following four characteristic phases: (1) rapid growth and exploitation (r); (2) conservation (K); (3) collapse or release ("creative destruction") (Ω); and (4) renewal and reorganization (α).

The Ω- and α- stages are brief periods of major change. In the Ω-stage, the system experiences a major disturbance or perturbation and some of the components or attributes of the ecosystem may be lost. The α-stage represents a period of reorganization and provides an opportunity for novelty to arise in the ecosystem. The r-stage represents a new phase or new trajectory in a well-defined basin of attraction. In the K-stage (or conservation stage), net growth slows, resources become increasingly locked up, and the system becomes increasingly interconnected, less flexible, and more vulnerable to external disturbances and shocks. Taken together, these four stages define an "attractor state." In the context of such an attractor state, building or maintaining resilience means preserving the entire attractor state (not just any one phase) and the ability of the ecosystem to cycle through all four states or phases, especially the release and reorganization phases.

Holling provides a number of examples of such attractor states in nature. One such example is the process of growth, accumulation, release, and recognition, for example, the fire, decay, and regrowth that occurs in forests. A patch of forest that has recently been razed by fire, if it still contains sufficient genetic material, will begin a new r-stage of exploitation, as early-succession forest species take root. The forest patch will gradually progress through stages of plant succession and growth in biomass as it transitions to the K-conservation stage; here the forest biomass is dominated by mature, so-called climax species. In a natural forest, biomass will accumulate to a "tipping point," where there is a sudden release or Ω-stage – a natural forest fire in a patch of forest – that burns the patch but not the entire forest. At the end of the quick release or Ω-stage, the system enters a period of reorganization, or the α-stage. In the reorganization or α-stage, there is opportunity for exotic species to invade the forest patch and modify the species composition through the "next cycle" of the attractor. If insufficient genetic material, or "history," exists in the system (i.e., the collapse is so severe that nothing is left for r-exploitation), then the system may "flip" to a new attractor state. Historically, forest management practices have acknowledged only the K-conservation stage and have focused on artificially maintaining the forest in this stage by suppressing natural forest fires. The end result of such a practice is that the forest ecosystem is set up for a much larger catastrophic crash, when a massive forest fire devastates not a patch of forest but thousands of hectares. A resilient forest ecosystem will contain many coexisting variable forest patches, each at a different stage of the overall attractor state. This variability will help preserve the overall ability of the forest ecosystem to maintain the global forest attractor state.

The ideas underlying Holling's Figure 8 are used for interpreting the data gathered in the case study that forms the basis for this chapter. We suggest that even though vulnerability and resilience are often portrayed as opposites in the literature on human security, natural disasters, and climate change adaptation, the two concepts are not opposites but complementary. We argue that vulnerability (V) can be reduced and adaptive capacity increased through a process of building resilience (R). These dimensions are featured in the results from the case study described in the following sections.

Case Study

The research for this chapter is founded on a case study of the Deerwood Soil and Water Management Association (DSWMA) as it undertook work within the South Tobacco Creek Watershed in Manitoba from 1985 to the present. DSWMA is a unique example of community-based management on the Prairies. Farmer-led since its inception over twenty years ago, and supported by all farmers in the watershed, the DSWMA was formed to address local concerns with soil erosion and flooding during extreme precipitation events, particularly the spring runoff and summer storms. Working in the South Tobacco Creek (STC)[1] Watershed, a 75-square-kilometre watershed bounded on the west by the Manitoba Escarpment and to the east by the Red River Basin, with a 183-metre (600-foot) elevation drop from west to east in just 11.3 kilometres (7 miles), the farmers began designing and building a network of twenty-six small check dams in the upper reaches of the watershed. Past research has demonstrated that the dams reduce peak flow runoff by as much as 90 percent. Equally important is the network of partnerships that the DSWMA has formed with over twenty other groups, including provincial and federal agencies and other local non-governmental organizations (NGOs).

The activities of the Deerwood group that have led to social and ecological resilience over the past twenty years can be depicted using the framework of the adaptive cycle based on Holling's Figure 8 diagram (see Figure 14.2). The DSWMA has undergone two major revolutions of the adaptive cycle, and may be in the process of a third reorganization. The first cycle began in 1980, with the founding of the DSWMA as a farmers' informal group. The first period of reorganization was from 1981 to 1984, when the group underwent transformation from an informal group into an official local organization, funded under a joint federal and provincial economic development program. The group was firmly organized and running as of

Figure 14.2

Adaptive cycles of the Deerwood Soil and Water Management Association

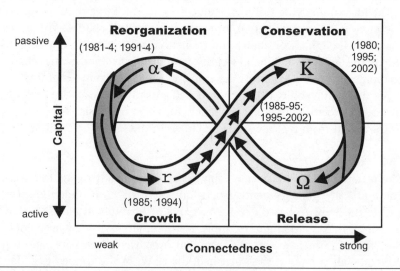

Source: After Gunderson and Holling (2002), with original data.

1985 and experienced a significant period of growth from 1985 to 1995, when it focused on on-farm watershed management activities such as dam building and promotion of zero tillage.

The loss of government funding after ten years, in 1995, marked the first significant crisis or release for Deerwood, and demonstrates the role that social-ecological resilience may play in carrying an organization through a successful reorganization process. Deerwood actually started reorganizing as early as 1991, in anticipation of losing government funding. The reorganization involved founding the South Tobacco Creek Pilot Project; creating the STC Steering Committee of experts; and shifting from on-farm activities to on-farm research. After this reorganization, the group enjoyed a second period of growth from 1994 to 2002, carrying out a variety of research in the STC under the auspices of the STC Pilot Project. In 2002, a second major funding crisis occurred, forcing the group to once again rethink its organization. While it is difficult to say for certain – as it is an ongoing process – Deerwood may now be in a second transformation stage, with the development of the Tobacco Creek Model Watershed project, which will, if successful, expand the DSWMA model all the way east to the Red River at Morris, Manitoba.

Key Vulnerabilities for the DSWMA

The term "vulnerability" can be used to characterize the crisis or crash that precipitates a release and reorganization phase in a resilient system. Here we focus briefly on the key vulnerabilities – namely, exposures and adaptive capacities – that existed when the DSWMA was first formed. Vulnerabilities were both biophysical and social in nature; in fact, it is nearly impossible to separate biophysical vulnerabilities from social ones, which demonstrates the tight nesting of the biophysical and the social. Farmers do not exist separately from their land but rather in an intimate, interwoven relationship.

We see this in several ways. The first is the reality of the geography and topography of the landscape where the original Deerwood members farm. The prominent and dominant landscape feature is the Manitoba Escarpment, formed out of the former banks of the ancient glacial Lake Agassi. The founding Deerwood members all farm on the top of the escarpment, on rolling hills and highly erodable soils. Layered on this physical landscape are two key features: (1) historical farming practices, including conventional tillage and summerfallow practices, informed by historical knowledge imported from Europe with the first settlers of the area (many of whom were the grandfathers and great-grandfathers of the original Deerwood members); and (2) historical land management practices, including clearing bush and draining wetlands, informed by the historical attitude towards water as a nuisance, or "get the water off the land as quickly as possible" and "clear as much land for agricultural production" as possible. These farming and land management practices combined with this particular landscape to create a very vulnerable situation, with low yields and high soil erosion and topsoil loss.

Another observation is that these biophysical vulnerabilities combined with some specific social vulnerabilities in the 1980s, including extremely high interest rates and record bankruptcy rates. These two forces of vulnerability working together – the biophysical and the social – set the stage for the first transformation of Deerwood from an informal local organization to a formal, government-funded and named "local organization."

Also important is the role that exposure to extreme events played at the time. Over the longer term – stretching back from the 1950s through the 1970s – the dominant extreme-event concern was too much water, especially during spring runoff and summer storms. The 1950s and 1960s were wet years, so there was an ongoing and vocal concern to do something to deal with flooding and related soil erosion. At the same time, the more immediate experience was with drought – the 1980s were very dry years on the Prairies, with significant droughts in 1980 and 1988-89 – and so farmers were motivated to do anything they could to try to improve yields and stop soil erosion at the time.

In addition to the previous set of exposures, there were also a number of key "adaptive capacities" that were part of the overall picture of vulnerability at the time of the emergence of Deerwood in 1980. All the founding Deerwood members were young farmers at the time, in their twenties and thirties, and most started farming between 1979 to 1982. They were young and very enthusiastic, with a "we can do anything" attitude, according to several of the members interviewed.

All the founding members also owned small mixed farms in very close proximity to each other – dotted along two or three side roads off the major highway passing through the area – within 10 or 15 minutes' driving distance. The small size of farm, from about 500 to 1,000 acres, translated into a relatively high density of local people in the area.

The group also had very strong "horizontal" social networks. They had all grown up together in the area, had gone to high school together, and had a common meeting area, a local seed plant that one of the founding members ran out of his yard. So they would often congregate there and chat informally. Strong vertical cross-scale linkages also existed. One founding member was on the local rural municipality council, another worked for the Manitoba Department of Agriculture, so both of these members had access to information and decision makers inside the government hierarchy.

Building Ecological Resilience
In the first adaptive cycle, from 1985 to 1995, the major and unique watershed management activity undertaken by Deerwood was the construction

of a network of twenty-six small dams in the upper reaches of the South Tobacco Creek Watershed. The group built three types of dams:

- dry dams, which do not store water but simply hold water back and slow the flow by forcing the water to flow through a small pipe 1 metre in diameter at the foot of the dam
- backflood dams, which are only 2 or 3 feet high and flood an area of a farmer's field and hold back the water for two or three weeks, until the farmer opens a gate and releases the water downstream
- multipurpose dams, which not only hold back the water but can store water for irrigation or cattle watering during the summer. These are fully drained in the fall to make the full capacity available to handle the spring runoff.

Besides building the network of small dams, Deerwood promoted a number of soil and water conservation activities to farmers in the watershed, including:

- conservation tillage, including zero and minimum tillage practices
- shelterbelt construction to prevent wind erosion and trap snow on fields
- grassed waterways, which involved the planting of grasses along the banks of waterways and in natural watercourses on the landscape to prevent soil erosion
- stabilization of natural gullies in fields through reshaping and planting of grasses
- forage planting, or taking marginal farmland out of annual production and sowing down to annual forages.

On the ecological side, it is evident that Deerwood's watershed management activities have reduced the impact of extreme weather events on the landscape. Prairie Farm Rehabilitation Administration (PFRA) studies document that the dam network reduced peak flows during extreme precipitation events by up to 25 percent. Figure 14.3 shows a rain event for the Steppler Dam in 1991, where the dam reduced peak flows by 90 percent.

Department of Fisheries and Oceans (DFO) Canada research shows that the dams hold back sediments and nutrients and act as artificial wetlands, improving water quality. The dams have also reduced flooding and soil erosion. Farmers have noticed a marked decrease in the turbidity of the creek water after summer storms. Increases in turbidity have been demonstrated to correlate with increases in water contamination and waterborne disease, so decreasing turbidity potentially contributes to improving water quality.

Building Social-Ecological Resilience

There are seven key features of a resilient social-ecological system (Olsson et al. 2004). The first feature is *leadership and trust*. At the local level, a core

Figure 14.3

Reduction of peak flows by the Steppler Dam during a June 1991 rain event

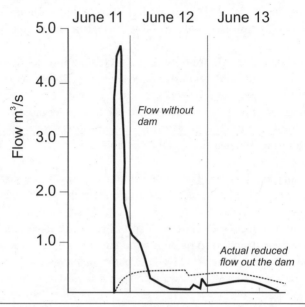

Source: Deerwood Soil and Water Management Association.

group of members provided strong leadership and was able to gain the trust of the larger group of Deerwood members. Several of these local members held key positions in other groups that exercised power (e.g., local and provincial governments). Leadership also came from a very well respected federal employee in the PFRA, who believed in what the Deerwood group wanted to do and committed to helping make it happen.

A second key feature of resilient social-ecological systems is *enabling legislation*. Deerwood was able to negotiate two key agreements with the province. The first was a licensing agreement that said that as long as the dams were under 4 metres high and held water on the farmer's own property only, Deerwood did not need a licence to build them. This flexibility gave the group the freedom to build dams. The second agreement was also critical for gaining the interest of a large number of farmers in having the dams built. Originally, Deerwood was told that the dams could only be dry dams, that is, they could not store any water. Most of the farmers were interested only if they could build a multipurpose dam and get a licence to hold water. Deerwood managed to negotiate an agreement with the province such that the dams could have a 50 percent storage capacity provision during the summer for on-farm water use, and then it agreed to drain the dams in the fall so they would be empty in the spring.

Access to flexible and significant funding is a third aspect of social-ecological resilience. Under the federal and provincial joint Economic and Regional Development Agreement, Deerwood received $359,000 in their first five years to build dams. In the individual interviews, a number of the Deerwood members stressed how important they felt the access to real, significant dollars was to their early success. The fact that they not only could plan activities but also had the funding to quickly implement their ideas gave them credibility in the eyes of the local members and helped them gain credibility with other partners.

A fourth key element of building resilient social-ecological systems is *monitoring*, which makes it possible to evaluate progress towards a goal. Through the South Tobacco Creek project, Deerwood has developed an extensive system for monitoring water flow and quality in the watershed. Monitoring has developed into a central activity for Deerwood and has contributed to its unique status; very few other watersheds in Canada have over ten years of baseline data on water flow and quality.

Social-ecological systems also enhance their resilience in a fifth way when they are able to *transfer information* effectively. The Deerwood group was able to do this with a Steering Committee in the early 1990s. The committee meets several times a year to coordinate the various research projects that the association is conducting. These meetings provide an important opportunity to merge local and expert knowledge; they also function as a driver in the ongoing adaptive management in the watershed. Members of the Steering Committee come from all four scales of partners, which helps them overcome their tendency not to collaborate. Some members mentioned that the only way they find out what is going on in other departments is through the committee meetings. Another mentioned that the Steering Committee allows them to focus on their common goal and leave behind interdepartmental battles and rivalries.

The Steering Committee also provides a venue for *combining information* from different sources, especially local knowledge with expert scientific knowledge. At the same time, it offers an arena for *collaborative learning*, as the group analyzes the results from various projects and learns together from its successes and failures. According to Olsson and colleagues (2004), these aspects constitute the sixth and seventh factors for building social-ecological resilience. As one member pointed out, the Steering Committee not only provides an excellent venue for collaboration but, because it is convened by the local Deerwood group, it provides a place for genuine community-driven science and learning (Funtowicz and Ravetz 1993; Lee 1993).

Conclusions: Policy Implications

A number of important policy implications have emerged from this work. In particular, the case study demonstrates that adaptive capacity for water resources management in an agricultural community requires:

- strong leadership not only at the local level but also across scales, at the provincial and federal levels. In particular, the importance of individuals at the federal and provincial levels committed and dedicated to the success of local community-based adaptation should be recognized.
- appropriate and enabling legislation that removes barriers to local communities' effectively implementing locally appropriate adaptation strategies. A one-size-fits-all approach should be avoided.
- real funding. The Deerwood case clearly demonstrates that adaptation cannot happen without a commitment by governments to genuinely fund adaptation initiatives. This is particularly the case where the benefit may be more of a public than a private good.
- accurate monitoring. It will be very difficult to understand future change without a clear understanding of our current baseline ecological conditions.
- innovative arenas for sharing information and mechanisms for facilitating the flow of information with and among all stakeholders across scales from the local to the national. To contribute successfully to an ongoing process of adaptation, this arena for collaboration needs to provide space for learning, experimentation, and innovation, and, as the Steering Committee does, provide a forum where the local community can interact with the scientific community in the true spirit of "engaged citizen science" or "post-normal" science.

The Deerwood experience also provides some very important lessons for future community-based adaptation to climate change in agriculture in rural Canada, at least in this context of the Manitoba Prairies.

When the researcher asked, "Could a community group like Deerwood emerge in the area today?" most of the farmers responded that they did not think so. They felt that Deerwood's future was aligned with what they saw as farming's future in general, which faces a number of significant challenges:

- A demographic shift in farming has been occurring. The farming population is fast aging, and the average age of farmers is now around 55. This contrasts with the average age in the Deerwood group when it got started, between 20 and 30.
- As farms are sold, the land is being consolidated into larger and larger farms. In this area, one particular cultural or religious group – the Hutterites – is buying up a lot of land. Hutterites now own about 50 percent of the farmland in the rural municipality of Thompson. The farmers who

mentioned this say that it is neither good nor bad; it is just a change, given that the Hutterite colonies, at least so far, have not demonstrated much interest in watershed management (they are clearing land with up to a 30 percent slope, laser-levelling the field, and using conventional tillage practices), and they appear uninterested in interacting with neighbours beyond their colony. This reduces the traditional social capital or networks that historically were so important to the founding of a community group like Deerwood.

- A community group like Deerwood relies on volunteers. With the changing demographics and cultural makeup, there are fewer and fewer young farmers, and those who are there are much more isolated and, the farmers believe, do not have the same sense of community that they had twenty years ago. This is potentially a significant barrier to future community group success.
- There is also at this time a lack of funding and of commitment and interest from the provincial and the federal governments regarding this type of community-based work. This is evidenced by the fact that although Deerwood is a demonstrated success and is held up nationally as a model of community-based watershed management, the group struggles to find meagre funds with which to operate.

Despite these challenges, we believe that the basic model of building resilience in response to vulnerability to extreme events includes elements that are useful for future adaptations.

Notes

1 Although small, the STC Watershed incorporates three distinct Prairie physiographic units (Uplands, Escarpment, and Lowlands), enabling the study of adaptive strategies appropriate to much of the Prairie landscape.

15

Household Access to Capital and Its Influence on Climate-Related Rural Population Change: Lessons from the Dust Bowl Years

Robert A. McLeman

It may at first appear anachronistic that the research presented in this chapter should be included in an edited volume on adaptation to climate change in Canadian agriculture. The empirical work on which the theoretical developments described here are based refers to another time and place, examining rural populations in 1930s Oklahoma and the effects of harsh climatic conditions on migration patterns seen there. Yet, at the workshop from which this volume arose, the relevance of an American example to the risks faced by Canadian farmers and farm communities today was never called into question. Quite the opposite: the "Dirty Thirties" were as bad for Canadian farmers as for their American counterparts. The memory of that unusually harsh period in agricultural history is etched into the collective memory of today's producers, and each one is well aware of the impacts a combination of harsh climatic conditions and low commodity prices can have on rural populations.

This chapter is not a reminiscence of those bad old days, nor is it an idle warning that what happened once could happen again (even if we all know that this might be true). Rather, the material presented follows an established method in climate change research, where historical analogues are used as a means of assessing and anticipating the future effects of climate change on human populations (Glantz 1991). My particular purpose here is to advance a theory of how climate change may affect rural population patterns, and to do so I examine how rural populations of eastern Oklahoma in the second half of the 1930s adapted, given the impacts of successive years of drought and flooding on such populations. One of the ways rural residents did so was by leaving the rural community altogether, many migrating to California in the fashion made famous by John Steinbeck's 1939 novel *The Grapes of Wrath*.

Such a migration was neither a random nor wholesale outpouring of people from rural areas. Rather, distinct groups within the rural population migrated out of eastern Oklahoma. Their vulnerability levels can be partly distinguished from those of the people who remained behind on the basis of their households' "adaptive capacity," understood in terms of their access to economic, social, and cultural capital. Such findings are consistent with recent developments in migration theory, and present an avenue for anticipating how rural populations may respond to future climate change, here in Canada and in similar, market-based agricultural economies. Before proceeding, it is worth making more explicit what is understood by the term "capital," and the way in which it is employed in the following case study.

The Concept of Capital

The concept of capital can be traced back to classical economic works such as those of Smith, Ricardo, and Marx. Classical economists used capital in the description and interpretation of various features of the production of economic goods, generally in reference to stocks, goods, or commodities beyond those required to meet basic human needs.

Various types or categorizations of capital have been identified by social scientists, some physical, some less tangible. Since as early as Adam Smith, for example, a distinction has been made between those forms of capital embodied in goods or the physical means of production of goods (which can be described as *economic capital*) and that capital which is embodied in people (Smith 1838). This latter form of capital, described by later social scientists as *human capital,* is manifested through the acquisition of skills, talents, and abilities that come through study, experience, or apprenticeship, and that increase the profitability of that person's labour.

In the case of agricultural production, *land* is treated by classical economists such as Ricardo (McCulloch 1852) as being a separate type of economic capital. Unlike other forms of capital, land is not movable, and therefore the person who controls it has a monopoly power over that space (Harvey 1973). Land is permanent (although its productive qualities may not be) and it can be put to multiple uses, so land ownership typically changes hands less frequently than other forms of capital. Moreover, and this becomes relevant to questions of population movements in the case study that follows, people must physically occupy some space; ownership of land confers upon the owner the right to its occupancy. Those not owning land must generally exchange some other form of capital in order to occupy a given space indefinitely.

Later social scientists have suggested that capital (*social capital*) exists in aspects of social structures and facilitates certain actions of people within the structure (Coleman 1988). These aspects are capital in that they are productive and can assist those with access in achieving ends that are not possible in its absence. Unlike human capital, social capital is not embodied in an individual but manifests itself in obligations, expectations, and trust between people; channels of information and communication; and norms and sanctions on behaviour. Social capital operates at a variety of levels or scales within society, and can be converted to other forms of capital. Social organizations, such as churches or organized recreational clubs, can facilitate the transmission and development of social capital. Social capital is not limited to groups at local scales but may operate at regional or interregional scales (Putnam 1995). Investigation of the role social capital may play in furthering economic well-being of communities has consequently emerged in international development studies in recent years.

Adding to those forms of capital already described, French sociologist Pierre Bourdieu introduced the idea of *cultural capital* (Bourdieu 1986). In some ways an extension of the earlier concept of human capital, cultural capital can be found, for example, in books and similar repositories of skills and knowledge, in the credentials obtained through formal academic or vocational training, and in the dispositions and well-being of individuals. As with other forms of capital, cultural capital can be transformed under certain conditions to another form (e.g., only those who have been admitted to the bar in a given province may practise law there and earn the associated salary; even lawyers who have trained elsewhere may not do so).

The total amount of capital and the different forms of it available to any given individual varies over space and time, implying that people's behaviour at a given point in time cannot be explained entirely by their particular capital endowment at that particular moment in time. To better explain their behaviour requires an understanding of their past conditions of existence (such as family origins and past capital endowments) and current social trajectory.

This very brief summary of the nature of capital and its role in social behaviour relates directly to the key problems being addressed here, namely, the untangling of the relationship between climate and migration. Responses to climate change can occur at a number of levels, from the individual or household level. They can then be deliberate or unintended and will vary according to time and place. Responses are also influenced by access to various forms of capital. Others have noted elsewhere that differentials in wealth between and within populations at different scales will affect the response options and decisions of populations exposed to the effects of climate change (Handmer et al. 1999). Adger (2001) has suggested that social capital may serve at the local level to differentiate levels of vulnerability to climate change among groups, and may help communities cope with extreme weather or climatic events.

In recent decades, capital in a variety of forms has been shown to influence migration behaviour, including economic capital (Mueser and Graves 1995), social capital (Massey and Espinosa 1997), cultural capital (Bauder 2003), religious capital (Myers 2000), and human capital (Larriviere and

Kroncke 2004), among others. Nee and Sanders (2001) used a framework that captured multiple forms of capital in explaining the choice of destinations among recent immigrants to the United States. If migration is considered in the context of the range of possible adaptive responses people exposed to adverse climatic conditions could employ, capital provides a bridge between climate change vulnerability and migration theory.

The following case study summarizes research conducted in Oklahoma and California in 2004, in which I used household access to capital in three broad forms – economic, social, and cultural – as a means of differentiating people who migrated out of a rural area heavily hit by multiple years of droughts and floods from those who remained behind. I suggest that such findings are not unique to the case study, and that differential household access to capital shapes and constrains a household's ability to employ migration as an adaptive strategy. The case study also suggests that use of capital in this fashion may be extended analytically to investigate adaptive capacity in a general sense. Lessons for those concerned about the ability of rural populations in Canada to adapt to future climate change follow.

Rural Eastern Oklahoma in the Dust Bowl Years
It is estimated that in the second half of the 1930s, up to 300,000 people migrated out of the southern Great Plains states of the US, the majority of them destined for California and other Pacific coast states. Roughly one-third of these migrants originated in the state of Oklahoma, and within this state, the greatest movement of people occurred within the eastern, central, and southern counties, where cotton and corn were the main crops (Figure 15.1) (Duncan 1943; Gregory 1989; McWilliams 1942; Stein 1973). This is not the semi-arid landscape one associates with the American Dust Bowl of the southern Great Plains or Canada's Palliser's Triangle. Rather, it is a rolling landscape furrowed by the Arkansas River and its tributaries, similar to the area around Riding Mountain in Manitoba or the farmlands of Eastern Ontario. Oklahomans refer to it as the "humid" or "green" part of the state. In the 1930s, this region supported a dense rural population, mostly living on small family farms.

Of the two main crops in the region, cotton and corn, cotton was grown for cash income. Cotton lent itself to cultivation in eastern Oklahoma for a number of reasons. During the First World War, cotton prices had reached record highs (this was also true of other commodities), sparking a surge in cotton production across the southern US. It was a crop with which the original farming settlers of eastern Oklahoma, many of whom came from southern states east of the Mississippi, were already familiar. Cotton has relatively low soil moisture requirements, and could therefore be grown on the hilly uplands of the region, where soils are thin and precipitation runs off quickly. Perhaps most importantly for eastern Oklahomans, reliable

Figure 15.1

Cotton- and corn-producing regions of Oklahoma in the 1930s

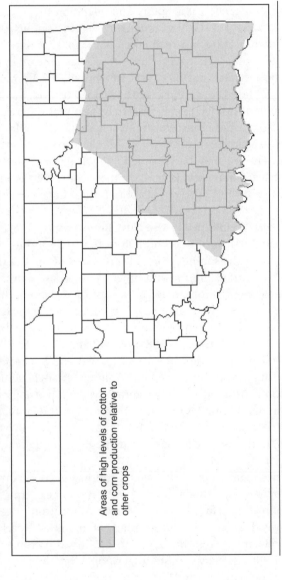

Areas of high levels of cotton and corn production relative to other crops

Source: After Ellsworth and Elliott (1929).

mechanical equipment for cotton production did not come into wide-spread use until the Second World War. Cotton could therefore still be grown profitably on small acreages, using draft animals and family labour, into the 1930s.

Because eastern Oklahoman farms had teams of draft animals, in addition to other livestock and poultry, significant acreages of corn were planted for local use as feed. Corn has higher soil moisture requirements than cotton, and so the distribution of crops at the farm level was such that corn was planted wherever precipitation runoff collected (referred to locally as "bottomland") and cotton was planted on slopes and higher ground. A vegetable garden for family use, a few hogs, and a coop of egg-laying hens completed the typical family farm.

Family farms of this description occupied roughly 80 percent of the rural landscape in a typical 1930s eastern Oklahoma cotton county like Sequoyah County, where my field research was conducted. The other 20 percent of the land was the most productive land in the county, and was occupied by corporate farms (Mayo 1940). Such farms were concentrated along the broad floodplain of the Arkansas River, where soils were deep and rich, and water was seldom scarce. These lands had once been operated by independent farmers, who had borrowed heavily during the boom years of the First World War to invest in equipment. In the 1920s, a rise in interest rates, combined with a postwar decline in commodity prices, led to the foreclosure of most independent farmers, and their lands and assets were taken over by business interests based in Fort Smith, Arkansas. Unlike the densely populated uplands, few people actually lived on these highly productive floodplains, where farm labour was performed by a small number of hired men using tractors. This land-use/land-settlement pattern is still evident today; the only areas of Sequoyah County where commercial crop production continues to occur on any scale are along the Arkansas River.

A trend that came to a head in eastern Oklahoma in the 1930s was a steady increase in the rates of farm tenancy, which in Sequoyah County represented 70 percent of all farms (Coleman and Hockley 1940). A number of factors accounted for this trend. Oklahoma was one of the last parts of the US west of the Mississippi to be opened up to homesteading. The lands of the state had originally been set aside in the 1820s for occupancy by Indian nations displaced by European settlers from their original lands east of the Mississippi. Much of Sequoyah County, for example, was settled by the Cherokee people, forced from their original lands in Georgia and from a subsequent reservation in Tennessee. With demand for homesteading lands still high but most western lands already homesteaded, the federal government changed its policy to allow non-Native settlement of Oklahoma. In a series of land runs beginning in 1889 and ending six years later, non-Native settlers poured into Oklahoma, quickly occupying any unclaimed lands.

From that point onward, farmland in Oklahoma could be obtained only through purchase or leasing.

In the first decades of the twentieth century, cotton could be produced profitably on small acreages using draft animals and family labour. This was not the case with other cash crops in the American West, such as grain crops, where large acreages and tractor-drawn equipment were already the norm. Consequently, a would-be farm family could enter into cotton farming with little economic capital. The lands of eastern Oklahoma, for the most part unattractive to mechanized agriculture, became increasingly attractive to the landless farmer once homesteading was no longer an option.

Several forms of tenancy arrangements emerged in eastern Oklahoma (Southern 1939). The most common arrangement was a crop-share agreement between tenant and landowner. Under such agreements, typically made with a handshake, the tenant would provide his own draft animals and equipment and seed. The landowner provided the land and a basic dwelling house. At harvest, the landowner received 40 percent of the cotton harvest and 60 percent of the corn (such agreements were thus typically known as 60-40s). Such agreements had to be renewed each year after harvest.

Poorer farmers who did not own draft animals entered into sharecropping arrangements similar to those in the American Southeast. In such arrangements, the tenant supplied his family's labour, the landowner almost everything else. The landowner received a much greater percentage of the harvest in return. Such arrangements were not especially common, representing only about 20 percent of all tenancy agreements. Similarly infrequent were agreements where farmers rented land for straight cash leases. Because tenancy agreements were renewed annually, there was a high turnover in farm occupancy, as much as one-third each year. While tenancy allowed farmers with limited economic means an entry into farming, it also meant that capital investment in farm improvements tended to be minimal (McDean 1978).

Beginning in 1930, demand for tenant farmland in eastern Oklahoma surged, and counties like Sequoyah experienced population increases in the first half of the decade as a result. The Stock Market crash in 1929 had a severe impact on railways and oil and mining companies, the main employment sectors in Oklahoma besides agriculture. Unemployed wage labourers from other parts of the state came to counties like Sequoyah, where the supply of tenant farms was high, for on a farm they could feed their families and hope to make a little money from cotton. At the same time, an extended period of drought had begun to take hold in the western, Dust Bowl parts of the state. With crop prices hovering at record lows, undercapitalized, indebted, and tenant farmers from that part of the state were forced off the land, many migrating to the humid east to start again.

In 1934, drought conditions spread straight across the state of Oklahoma into the eastern counties (Figure 15.2). Cotton and corn crops failed almost

Figure 15.2

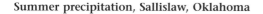

Summer precipitation, Sallislaw, Oklahoma

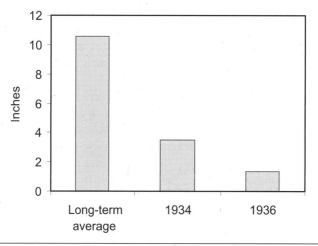

Source: United States Department of Agriculture Weather Bureau (1934, 1936).

completely. The following year, extreme summer precipitation events caused widespread flooding and destroyed large areas of cropland in eastern Oklahoma. The year 1936 saw drought conditions even more severe than those of 1934, and cotton and corn crop failure was universal in eastern Oklahoma.

Cotton prices were already at record lows in 1934 when these successive years of crop failure began, and living standards on Oklahoma farms were already well below those in other states (McDean 1978). Local observers began predicting an exodus from Oklahoma farm country, and by 1936, large-scale out-migration was well underway (Duncan 1935; McMillan 1936; Taeuber and Hoffman 1937).

Impacts of Climatic Conditions

Regardless of socio-economic status, virtually every eastern Oklahoman household suffered a loss of income as a result of the harsh climatic conditions of the mid-1930s. Successive years of cotton failure wiped out the cash economy, and barter became the main form of economic transaction. While barter and short-term credit at stores and suppliers had always been used as a means of getting through short-term financial problems, it could not sustain the rural economy over multiple crop years. The majority of farmers did not own the land they worked, meaning that, even if they could find a bank still willing to lend to farmers, of which there were few, they did not have collateral for a mortgage. Hence, farmers experienced a dramatic erosion of their economic capital, with most having no means of offsetting it.

While farmers who owned their land tended to employ the same methods and grow the same crops as tenant farmers, often on adjacent parcels of land, a gulf rapidly emerged between those who owned land and those who did not. Few tenant farmers had multiple-year agreements: after a single crop failure, their tenancy agreement might not be, and often was not, renewed. Even after a complete cash-crop failure, a landowning farm family could at least get by on its own produce, and remain in its home through to the following planting season.

Despite repeated crop failure, demand for tenant farmland continued to remain strong in eastern Oklahoma, so bad were conditions elsewhere in the state. This enabled landowners to begin asking for "privilege money," an up-front cash payment from tenants in addition to the standard crop-share arrangements. At the same time that demand was running strong, federal agricultural policies were providing landowners with incentives to withdraw their land from production. The Agricultural Adjustment Act (AAA), which took various forms during the decade, was designed to stabilize low and falling commodity prices by paying farmers to reduce acreages of many crops, including cotton, and to reduce the number of cattle. It was not implemented as a measure to help farmers cope with crop failure. With repeated crop failures, and consequently little revenue from their crop-share tenants, it made economic sense for some landowners to evict their tenants and collect payments under the AAA. While the AAA undoubtedly helped many landowning farmers remain solvent through the 1930s, it also served to reduce the amount of land accessible to tenant farmers and to help scarce cash flow from tenants to landowners in the form of privilege money.

As cash became scarce, social capital became increasingly valuable in the rural community. Obtaining short-term credit from a local merchant is one typical example of social capital; a family had to be known to a merchant, and known to be "good" for the money, before such credit would be extended. Other manifestations of valuable social capital included young couples sharing a residence or farm to reduce their expenses, and extended families and close neighbours helping one another to perform labour-intensive work such as slaughtering hogs, to reduce the need for hiring casual labour.

Landowners typically enjoyed greater social capital within the local area than did tenants. As noted earlier, tenant farms had a high rate of turnover, even in good years, and tenants had no right of residency on a particular farm beyond the current crop year. Consequently, as a group they had less opportunity to form lasting ties to any one specific place within the region. Landowning farmers tended to be older on average than tenant farmers, and their farms typically passed from one generation to the next.

They were therefore able to establish networks and connections with other residents through institutions such as schools, churches, and community groups.

Financial benefits of such connections were realized with the implementation of the Works Progress Administration (WPA), a New Deal program designed to stimulate economic growth and employment through federal funding of local infrastructure projects in those parts of the country hit hardest by the Depression, including eastern Oklahoma. While the funding of such projects was federal, the administration of local projects was transferred to local officials. To get a job on a WPA project, or to get a contract to provide draft animals for use on WPA projects, one typically had to have connections with local administrators, since the number of people seeking such jobs far outstripped the number available.

In the mid-1930s, there was little demand in the regional labour market for the cultural capital eastern Oklahoma's farmers had developed through their agricultural practices. Earning a living as a crop-share tenant farmer in Oklahoma's cotton belt required a young family physically able to work for long hours, and the head of the family needed to acquire skills in managing draft animals. The growing of cotton was labour-intensive at three distinct points in the growing season: planting, chopping (the thinning of young plants), and harvest. The collection of cotton bolls was a task that was not only physically demanding but done efficiently only by acquiring a knack for detaching it from its thorny surroundings. Oklahoman farmers had often supplemented their farm income by journeying between growing seasons to neighbouring states to pick cotton or other crops. This tactic had often helped them get through a bad harvest, but in neighbouring states and agricultural regions, there were more people seeking casual agricultural work in the 1930s than there was work available. One government official testified that in 1939, the agricultural region along the Rio Grande in Texas, a traditional destination for seasonal workers from Oklahoma, was the scene of thousands of starving people vainly seeking work picking peas (Bond et al. 1940).

The tenant farmer rarely had the opportunity to develop any cultural capital other than that which suited him or her to cotton growing. Schools in rural Oklahoma were few and poorly resourced. Public education was funded primarily through property taxes, and landowners, who represented a minority of the rural population in eastern Oklahoma, had little incentive to support tax increases to improve educational opportunities for the majority. When cotton farming failed to provide a living, and when agricultural workers were not needed in neighbouring regions, the tenant farmer could turn only to manual labour jobs with railroads, mines, or oil companies as alternative occupations. Such jobs were scarce in the mid-1930s.

In summary, the harsh climatic conditions of the 1930s imposed hardships on all members of the rural population of eastern Oklahoma, but some groups within that population were hit much harder than others. A gulf emerged between the capital endowments of landowning and tenant farm households. High turnover of tenant farms and an influx of farm seekers from other areas increased the difficulty and cost of obtaining a renewed farm tenancy agreement, while New Deal programs offered the greatest benefits to those who owned land and/or who had well-established social connections with local government administrators. It is no surprise that the greatest percentage of migrants to California from rural eastern Oklahoma were drawn from the landless part of the population.

Role of Capital in Facilitating Migration to California

The migration out of eastern Oklahoma's cotton counties was neither wholesale nor random. In his classic study of the migration, historian James Gregory (1989) observed that Oklahoman migrants to California included a particularly high concentration of young married couples with children, and that many came because they had relatives who had settled in California previously. People from rural areas of Oklahoma tended to settle in rural parts of California, particularly in farm areas in the Great Central Valley such as Kern County, in the southern San Joaquin Valley. Drawing upon a range of sources, including scholarly research, autobiographies, oral histories, government reports, newspaper accounts, and interviews with members of migrant families, local authorities, and residents of Sequoyah and Kern Counties who witnessed the migration first-hand, I sought to develop a detailed understanding of the characteristics that distinguished migrant households from non-migrant households. In particular, I sought to identify and compare the endowments of economic, social, and cultural capital of migrant and non-migrant households.

While capital in its multiple forms has been used in research elsewhere as a basis for analyzing migration behaviour (see discussion above), to my knowledge this is the first attempt to utilize such an approach to investigate climate-related migration. Its use here develops considerable insights into the migration behaviour of Oklahomans during the 1930s, uncovering clear distinctions between groups who remained in eastern Oklahoma through the 1930s and those who migrated to California.

While tenant farmers were the main source group of rural Oklahoman migrants to rural California, patterns of capital endowments are identifiable that characterize those who migrated and distinguish them from those households that did not. While Oklahoman migrants were considerably poorer than established Californian residents, they were generally not completely destitute when they left Oklahoma. To get to California, most had sold off their possessions, draft animals, and farming equipment, and had purchased an automobile if they did not already own one. This is not entirely surprising; migration over long distances is not without its costs.

Social capital at a different geographic scale played a considerable role in facilitating migration. Migrants who left eastern Oklahoma for California did not generally do so on the basis of speculation alone; rather, they were following in the path of relatives, friends, or neighbours who had gone before them. By the 1920s, a sizable population of expatriate Oklahomans had already settled in southern California. In the oral histories and autobiographical accounts of 1930s Oklahoman migrants, one finds almost invariably that the migrant had a close relative who encouraged him or her to join them in California. These social connections between Oklahoma and California served three key purposes. First, information about employment opportunities and migration advice was transmitted in this way. Second, relatives in California could often be counted upon to assist new arrivals in finding accommodation and getting introduced to potential employers. Third, strong social ties within the Oklahoman community in California helped them maintain themselves in an unwelcoming and sometimes openly hostile environment. I devote extra attention to this third factor as it helped make the migration from Oklahoma to California a permanent change of residence, unlike past patterns, where Oklahomans had often migrated to other states on a temporary basis to find work, returning home for planting time.

The Great Central Valley of California was and still is a highly productive agricultural region. During the 1930s, tens of thousands of Oklahomans settled in the southern part of the Central Valley agricultural region, especially in Kern County, in the southern San Joaquin Valley. Unlike eastern Oklahoma, rain-fed agriculture is not possible in the arid conditions of the southern San Joaquin. While agriculture was still undergoing considerable growth and expansion in the area in the 1930s, the operation of a farm was possible only with sufficient financial capital to invest in irrigation. Although many Oklahomans who set out for California did so with the hope of someday acquiring their own farms, this was well beyond the means of most.

There was at that time, however, a significant demand in California for agricultural labourers with the skills that eastern Oklahomans tended to possess. Cotton was a major crop in the San Joaquin and, as in Oklahoma, the work was typically done with draft animals and large quantities of manual labour at chopping and harvest time. Such skills did not exist in the local California labour market. The San Joaquin's commercial growers were then (as now) heavily dependent on imported labour from Mexico to perform farm tasks at key times in the growing season. Such workers moved from farm to farm, following the work schedules of the various crops, and returned to Mexico when the harvests were all finished. With the onset of the Depression, however, the federal government placed restrictions on the number of foreign workers allowed to enter the US each year, reducing the size of the workforce available to Californian growers. The influx of Oklahomans, with their familiarity with cotton and willingness to perform such physically demanding work, was a timely development for the growers.

Oklahomans differed from the Mexican workers they replaced. Unlike Mexicans, Oklahomans did not leave the San Joaquin after the harvests were finished, but stayed and sought full-time work. They enrolled their children in local schools and demanded housing and health care, services that had not been provided to Mexicans. They demanded better wages, which sometimes brought them into direct and violent conflict with the growers. Some established residents of the valley were openly hostile to the newcomers, not welcoming them into their churches, and local government administrators often attempted to deny Oklahoman migrants access to basic services. The state lengthened the residency requirements for eligibility for social assistance benefits, even though the evidence is that only a minority of Oklahoman migrants resorted to it, and even then only in months when agricultural work was scarce.

In one notorious incident referred to as the "bum blockade," the sheriff of Los Angeles County, another favourite destination of Oklahomans, sent police officers to the California/Arizona border to harass and discourage new arrivals. Along westbound Route 66 heading out of Oklahoma, the state of California posted billboards telling jobseekers that they were not wanted in California. Steinbeck's *Grapes of Wrath* was published in 1939, drawing nationwide attention to the plight of California's Oklahoman migrants. It was banned from public libraries in Kern County, and a locally powerful individual in California's cotton industry symbolically burned a copy of the book in the streets of Bakersfield, the county's largest city.   .

In the face of this hardship, Oklahoman migrants pulled together, building their own networks and forming their own social capital (Stanley 1992). The administrative reports from the Arvin (CA) Federal Migrant Labor Camp, home to hundreds of migrants from rural Oklahoma and the basis for Steinbeck's Weedpatch Camp in *Grapes of Wrath,* contain countless references to acts of kindness and support that migrants performed for one another (Collins 1936). These included such simple things as collecting bus fare for someone's relative who went back to Oklahoma for a visit and got stranded, or women minding one another's children when someone needed to be away on an errand. Established Oklahomans would vouch for new arrivals to get hired for their first job, and information about which farms were hiring flowed through migrant networks.

In summary, the cultural capital they possessed enabled rural Oklahomans to quickly find work in rural California. A small amount of economic capital (from the sale of workstock and farm equipment) combined with social capital (in the form of family connections with people already established in California) helped facilitate the journey there and assist migrants in settling. The formation of additional social capital within the Oklahoman community in California enabled migrants to maintain their toehold there in spite of the poor wages and adverse conditions.

Although many Oklahoman tenant farmers had the cultural capital to work in California, not all had the economic or social capital to facilitate their migration there. What happened to those displaced from their farms and homes in rural eastern Oklahoma who lacked the capital to leave the region? Empirical evidence of their fate is difficult to uncover, but autobiographies and interviews report that throughout eastern Oklahoma, makeshift shacks began to appear on unclaimed land, under bridges, along railroad tracks, and along the banks of rivers and ditches. Large squatter camps formed on the outskirts of cities, a particularly large one being located on the edge of the Oklahoma City garbage dump. The population of these terrible places was most likely drawn from that group.

General Lessons from the Case Study

It is important to be cautious when using a single historical case study or analogue to anticipate the behaviour of other systems at other places and times (Jamieson 1988). The agricultural patterns, practices, and policies of 1930s Oklahoma no longer exist today. Cotton is no longer grown in eastern Oklahoma; the increasing costs of controlling boll weevil infestations and the widespread adoption of mechanized cotton-harvesting equipment made its production in that region uneconomical by the 1950s. A typical family farm in Sequoyah County today has a few dozen head of cattle, some pasture, and parents who earn most of their income off-farm. For those few full-time farmers who remain, there are today a wider range of technological options, insurance plans, and farm support programs than existed in the 1930s. Although there are still a great many horses in Sequoyah, these are not draft animals but racing stock, and the income they generate depends on their success at the track and not their strength in the field. Marijuana grown along lakes and watercourses in more remote areas may well be the most valuable cash crop, and some former sharecropper cabins in remoter areas are now home to illegal crystal methamphetamine production. The same coincidence of adverse climatic conditions, low commodity prices, and general economic recession experienced in the 1930s no longer produces the same effects in terms of population movements. Why, then, should inferences be drawn from this case study and applied to future climate change adaptation research in rural Canada?

The empirical research described here was undertaken to assess and further develop a theory of climate-related migration that was already conceptualized (McLeman and Smit 2006). It did not give rise to a capital-based theory of the climate/migration relationship; rather, it gives credence to it. The use of capital as a means of interpreting human migration behaviour has been done elsewhere for other types of migration events. The extension of this approach to this particular climate-related migration event and the results that came from doing so are consistent with applications elsewhere. In other words, the case investigated here is an early demonstration of how a theoretical approach that analyzes and incorporates household access to capital in all its forms may uncover how rural population patterns change in response to climatic stimuli. Such changes can be seen and interpreted in the context of vulnerability and adaptation to climate change, as is explored in greater detail by other authors in this volume. In looking for ways to identify, describe, and ultimately enhance the ability of Canadian rural areas to adapt to climate change and its impacts, this research project suggests that household access to capital may play an influential role.

As shown by other authors in Part 2 of this volume, the impacts of climate change are not expected to be uniform across rural communities, nor are they expected to be randomly distributed. Different groups within the rural community are expected to experience the impacts (or opportunities) of climate change in different ways. In looking for ways to uncover such differential experiences, it can be seen from the case of eastern Oklahoma that climatic conditions may have immediate and readily observable impacts on household endowments of capital. The most obvious of these is a decline in economic capital, which is to be expected in an economic sector like agriculture where favourable climatic conditions are a direct production input. It can also be seen, however, that with successive years of harsh climatic conditions, the value of social capital and cultural capital also changed significantly.

This theory also suggests that the capacity of households to adapt to such conditions is influenced or constrained by its access to capital in its various forms. In the Oklahoma case, those families who owned farms were better able to get through this period of unusually harsh climatic conditions than those who had some other form of tenure. This was not the only decisive factor, however. The ability to harness social capital – within the local region in the case of those who remained on the land, and outside the region in the case of those who migrated – became crucial as money became scarce. For those who went to California, social capital provided the means of making the geographic leap from the southern Plains states, where the value of their cultural capital was depressed, to another agricultural region, where such cultural capital was still in demand.

Lessons for Rural Population Change Research in Canada

For many years now, there has been concern for the vulnerability of Canada's rural areas. The closure of schools, the withdrawal of health care services, and the end of traditional economic patterns (e.g., where farms and rural businesses are passed on from one generation to the next) are familiar refrains in many parts of the countryside. One of the manifestations of this environment is a seemingly inexorable outflow of people from rural areas. Climate change and its possible effects are just one addition to the many forces of change confronting rural populations. The question therefore becomes, how may climate change influence existing patterns of population change in rural Canada?

Looking once more at the case study, the harsh climatic conditions of the 1930s served to accelerate the outflow of a particular portion of the rural population from eastern Oklahoma. Was this outflow inevitable? Would it have occurred anyway, but perhaps over a longer period of time? Even with hindsight, it is difficult to say, since the people who were forced off the land and who left the region – young families with skilled, hardworking parents – are the human capital on which a vital and viable community depends for its future development, whether in the countryside or elsewhere. Once this segment of the rural population departed, the socio-economic trajectory of the population that remained was changed irreversibly. What alternative futures the region might have had had climatic conditions been different in the 1930s can only be guessed at.

And so it is when we look to how climate change may affect rural populations in Canada. How capital in its different forms is distributed among rural households may influence their adaptive capacity and hence future socio-economic trajectories under climate change. Therefore, identification of the nature, distribution, and valuation of capital in all its forms in a given rural population, combined with investigation of how the likely manifestations of climate change may affect those capital endowments, may provide insights into the ways in which members of that population are vulnerable. Doing so may also shed light on what attributes may help or constrain members of the rural population in adapting to the effects of climate change, and identify those members most likely to adapt by out-migration.

I conclude with a final observation about the case of rural eastern Oklahoma during the 1930s that reinforces points made in Chapters 8, 9, and 11, which emphasize the importance of contextualizing responses to climate and weather stress. One of the features making this Dust Bowl case study informative is that challenging climatic conditions were not the only adversities confronting the rural population. The rural population also faced a steady erosion in commodity prices, increased competition with producers in other states and countries, ongoing intensification in agricultural production at a broad scale, and a severe general economic recession that limited the ability to find off-farm employment. At the time of this writing, Canada's rural communities are grappling with many of these same forces, as well as new ones, such as the closure of the American border to cattle exports because of the presence of bovine spongiform encephalopathy (BSE) in some herds. Rural people are resilient, and were it simply climate change on its own confronting them, there is every reason to believe that they could, and would, adapt and respond successfully. It is when confronted with a host of adversities simultaneously that their ability to adapt successfully to climate change may be pushed beyond its limits.

Part 5:
Conclusions

Part 5 summarizes what has been accomplished and what needs to be addressed to improve the information base for representatives of the agri-food sector and policy makers concerned about climate change adaptation for Canadian agriculture. Chapter 16 offers views from representatives of those two groups regarding the importance of climate change adaptation and the government's role in bolstering it through research and policy. Chapter 17 is a brief overview of this volume; it concludes with a summary of research topics and policy areas that need attention.

16
Policy Implications: Panellists' Comments
Ellen Wall, Barry Smit, and Johanna Wandel

Several authors of the chapters in this volume point out the important role policy plays in influencing producers' adaptive capacity for managing risks associated with climate change and variability. David Sauchyn (Chapter 6) notes that historically, settlement in the Prairies and its reliance on agriculture required many adjustments of practices and policy in response to climatic, social, and institutional crises. The highly adaptive capacity of the sector could not have been attained without government policy at many levels. Current conditions in the Prairies continue to require ongoing supportive government policy and programs. But, as Henry Venema (Chapter 9) and Harry Diaz and David Gauthier (Chapter 10) indicate, agricultural policy related to climate change has major implications for sustainability that must be considered if it is to be truly effective.

Michael Brklacich and colleagues (Chapter 2) note that Canadian agri-food policy has matured during the past twenty years primarily in response to a growing awareness of influences from a broader set of factors originating within and from outside the sector. The authors suggest that agricultural policy for climate change needs to continue to broaden the context and incorporate an increasingly dynamic and complex environment. They also recommend that policy consider the agri-food system's vulnerability as a property of the system itself and focus on socio-economic factors. Ben Bradshaw develops this theme further in Chapter 8, and points out that producers' future responses to climate risks and opportunities will probably mirror past responses. Policy makers need to keep in mind that producers have historically adapted to risks and opportunities by expanding, specializing, intensifying, and integrating their operations, not to mention seeking alternative income sources. Thus, according to Bradshaw, successful agricultural policy designed to build adaptive capacity for climate change and variability must take into account this reality. Christopher Bryant and colleagues concur in their discussion of Québec agriculture in Chapter 11. They note how the focus of climate change adaptation research in that province has broadened to include policy implications, both in relation to the relevant government agencies and to other players who have a stake in farming and who can influence directions through their own actions, programs, and policies, such as the Union des producteurs agricoles du Québec (UPA) and the Financière agricole du Québec (FAQ).

Denise Neilsen and colleagues (Chapter 7) demonstrate the value of incorporating producers' responses into water resource management when their regions experience climate- and weather-related moisture stress. In their case study, it is evident that several levels of government and institutions need to work together to ensure that programs are successful. Chapter 14 includes an additional example of the multitude of players involved in developing and implementing sound programs and policy for ensuring that producers have the adaptive capacity to handle the impacts of climate change and variability. In that chapter, Cynthia Neudoerffer and David Waltner-Toews point out the role of enabling legislation and access to funds in building resilience (and therefore adaptive capacity) among a group of agricultural producers working on water management projects. Farmers' experience in the mid-1980s enables them to handle current water resource challenges better than they might otherwise be able to. Going further back in agricultural history and looking south, Robert McLeman (Chapter 15) identifies agricultural policy as one of the key elements in migration patterns during the Dust Bowl era in the United States. Among the points he makes is how particular initiatives (for example, the Agricultural Adjustment Act), despite their intended goals of assistance, ended up producing outcomes that were negative for tenant farmers working the land.

More detailed analyses of producers' perceptions of risks (climate and otherwise) and options for managing them are offered by Suzanne Belliveau and colleagues (Chapter 12) and Susanna Reid and colleagues (Chapter 13). Information from the latter study points out the importance of income stabilization programs for producers and how climate and weather impacts are woven into market and economic risk factors. This finding is echoed in information from the Okanagan apple and grape producers provided in Chapter 12. In that case, the use of crop insurance and alternative varieties is a vital part of risk management. Other aspects of agricultural policy may not be that helpful when climatic and weather conditions are taken into account, however. As the authors note, some programs that encourage shifting to new strains and varieties may be leaving producers more susceptible to climate/weather risks.

Many of the policy issues raised throughout this book are supported by producers' views on what is needed for effective climate change adaptation policy (see Table 16.1). They note that just as climate risk is not managed in isolation from other risks, so too effective climate risk policy is not likely to be developed in isolation from other government initiatives affecting the

Table 16.1

Summary of risk management policy challenges, solutions, and recommendations relevant for climate change as identified by producers

Challenge	Specifics	Solution	Recommendation for policy
Economic variability	Variation in: • income • interest rates • energy costs • dollar value	Income stabilization	The main goal for agricultural policy should be agri-food sector stability. AAFC and provincial ministries of agriculture should ensure that the outcomes for the agri-food sector are considered when other ministries develop policy and programs that affect it. Income stabilization programs must be adequate for future climate and weather risks.
Sector variability	Variation in conditions and requirements: • across commodities • across regions • across types of farming systems	A "one-size-fits-all" solution is not possible. Flexible policies/programs that lead to equitable results.	Ensure that the diversity of conditions, needs, and expectations for all sectors/regions are taken into account in policy and program development.
Mainstreaming	Adaptation to climate risks must be considered in light of other business risk strategies. Farming systems management is highly integrated.	Identify opportunities for integration into existing strategies. Identify potential barriers to integration and uptake. Become aware of real farm experiences.	Substantial support for research is needed; it must feature a producer perspective and "whole-farm context." Research must include assessments of barriers to adaptation, including policy/program environment. Include climate change adaptation in the Agricultural Policy Framework; belongs directly in Business Risk Management but also relevant for other "pillars" (Environment, Food Safety, Innovation, and Renewal).

▲

▼ *Table 16.1*

Challenge	Specifics	Solution	Recommendation for policy
Barriers to adaptation	Some adaptation options for climate risk pose challenges for farming community: • additional costs to producers • genetically engineered solutions compromising marketing products • conflicts with existing policy	Research needed to identify: • adaptation costs/benefits • implications of genetic engineering technology • potential conflicts and ways to make them complementary	Support research that will: • provide long term and in-depth assessments • assess costs and benefits of climate risk adaptation options Develop policy and programs based on research findings.
Adequate support	Some options require improved resources: • technology lagging (e.g., weather fore-casting needs to be more reliable) • knowledge transfer and financial support (incentives) needed to encourage effective risk management	Improved product development for "technological" adaptation options (e.g, weather and climate forecasting). View farm management practices in light of climate adaptation options.	Re-establish research and extension services that work directly with producers. Establish climate change adaptation on "on-farm" demonstration sites.

Communication	Information about climate change risks is not always consistent or reliable. Insights from producers are not always recognized.	Improved resources for generating information. Enhanced "extension" services. Place more value on producers' knowledge.	Ensure that information from government is well supported through research and presented in useful formats. Require producer representation on research and policy development teams.
Enhancing capacity	Farming community needs more capacity to manage risks. Public image of agri-food sector can be one of "neediness."	Look at past examples that worked (e.g., need for new grape varieties resulted in successful collaboration between industry and government) Initiatives that reward sound management.	Work collaboratively with producers to ensure relevance of potential solutions. Aim for policy environment that provides assistance while promoting producers' independence.

Source: C-CIARN Agriculture (2004).

Canadian agri-food sector. Much work needs to be done to understand agricultural production from an operating farm perspective or "real-farm experience" so that practical expectations can inform policy and programs and contribute to useful assessments of barriers and opportunities.

Examples of potential problems for producers managing climate risk include uncertainty in the value of climate data, apprehension about the public acceptance of recommended technology (e.g., genetically modified organisms), and potential conflicts among programs and policies. Research support, effective technology transfer, and more collaboration among different government departments and ministries (as well as across federal, provincial, and, in some cases, local, lines) will go a long way in helping to generate policies and programs conducive for climate adaptation in the agri-food sector.

The seven types of challenges presented in Table 16.1 (economic variability, sector variability, mainstreaming, barriers to adaptation, adequate support, communication, and enhancing capacity) can be grouped into three broad issues to which producers and policy representatives have offered insights. These topics (variability/uncertainty, capacity, and adaptation processes) form the basis for the presentation of comments (in the form of verbatim quotes) from producers and policy representatives who took part in the Canadian Climate Impacts and Adaptation Research Network for Agriculture (C-CIARN Agriculture) workshop in Edmonton on 17 February 2005.

Variability/Uncertainty
As noted in Table 16.1, stability in all aspects of farming, including climatic and weather conditions, provides an ideal operating environment. Seven panel members made comments relevant to variability and uncertainty. Comments from producers are presented first, followed by comments from panellists working in the policy arena.

Bruce Beattie, Alberta dairy farmer representing the Alberta Environmentally Sustainable Agriculture Council:
   I think there is a bit of confusion between weather and climate; I can deal with weather because next year the weather will be different but, if it's climate that is going to change, then it is a whole different story. If you punch me in the side once or twice that's fine, but if you keep doing it for a long time that's different – it's going to start hurting.

Nigel Weber, Alberta family farmer (mixed production) with the National Farmers Union:
   If, because of climate change, we don't know what the relative economics of grain production versus livestock production will be in a given

area of the country, does it make sense to replace adaptable, mixed-production, family farms with corporate specialized producers? To put it differently, if we transfer hog production to corporate mega-barns and if climate change makes that industrial hog production impossible in the future – either through restrictions on water use or feed shortages – will those corporate producers adapt and begin growing crops or raising dairy cows? Most probably, those corporations will simply leave. As variability and uncertainty grows, the existence of adaptable, family farm agriculture becomes increasingly important.

Farmers are basically at the front lines of adaptation because they are the ones who really see and feel the effects because of their close relationship with nature. When there is a drought or a hailstorm – farmers are the ones who really feel that.

Don McCabe, Ontario cash crop producer with the Soil Conservation Council of Canada:

Don't expect anybody in Corntassle, Ontario, or Backslash, Newfoundland, to worry about it tomorrow if you are going to tell them the climate is changing. In order to be ready for where we are going to be tomorrow – to pick up the grasshoppers that a farmer is going to have in Alberta or the bugs that I am going to have coming in from Louisville, Kentucky to southern Ontario – we need to maintain capacity for wheat breeding, canola breeding and all the rest of it.

Bruce Burnett, Canadian Wheat Board:

The role of multiple stressors is very important; you can't deal only with internal factors. The world markets are extremely dynamic and respond to even small shocks. Economic forces will create a wholly different environment so we have to be very, very careful when using time scales, especially when policy can come in and change things quite rapidly.

Producer in the workshop audience:

At the production level, dealing with variability along the same general trends of climate change is probably the biggest issue a producer has to react to. Indeed, in terms of future planning, if variability is so great that no single plan will work, then it is extremely important to describe variability in terms of climate and the impact that that has on production, and then change the stability with which production practices can react to it.

If variability is so great that it is just a waste of time to plan ahead 15-20 years, then we need to spend some more time thinking about how we actually DO plan for variability under these circumstances.

Alan Stewart, Prairie Farm Rehabilitation Administration:
  Uncertainty and variability are directly linked to the longer term.
  The likelihood of certainty increases with the shorter term. Immediate
  concerns (of which seasonal weather/climatic condition is only one) are
  far more tangible. Predictions of a hotter, drier condition in the future
  or of increased variability in weather are not specific enough for making
  investment decisions and taking action. Thus, climate change is on the
  radar but the object isn't near enough for identification and to allow
  appropriate action to be taken.
    There is also variability or diversity in the types of communities
  involved in climate change adaptation and given uncertainty about the
  impacts of climate change on agriculture it is impossible at this stage to
  define the specific changes and investments required. Community-
  based approaches to adaptation must recognize that there are a very
  wide range of community types, and as many interpretations of
  "community" as there are interpreters.

Brian Abrahamson, retired from the Prairie Farm Rehabilitation Administration:
  There is a need to have a common understanding about the time period
  for which adaptation is targeted. The critical time period for producers,
  policy makers, industry, and researchers differs. As well the specific
  adaptation activity differs with each stakeholder.

Table 16.2

Suggested responsibilities and critical time frames for adaptation to climate change

		Responsibility[2]			
Adaptation activity	Critical time period (measured from present)	Producer	Industry	Government	Research
On-farm management	3-7 years[1]	●	○	○	○.
Agricultural policy	5-10 years	○	○	●	○
Agricultural infrastructure	30 years	○	●	●	○
Climate change adaptation research	50 years	○	○	○	●

Notes:
1 Annual crops and perennial forage (excludes orchards).
2 ● = lead; ○ = support.

For producers, current variability is the primary threat and the one on which they will focus their adaptation efforts. Table 16.2 summarizes the diversity among stakeholders in approaching climate change adaptation.

Nancy Lease, Quebec Ministry of Agriculture, Fisheries and Food:
A long-term vision for the sector is essential to allow for the long-term planning to support the healthy functioning and the adaptation of the sector in a global setting. A proper appreciation of the challenges represented by climate change, as well as by other issues such as environmental sustainability and international competition, is needed in order to guide development of longer-term policy for agriculture.

Capacity

It is a truism to state that the capacity to adapt to climate change depends fundamentally on a variety of resources. Whether at the level of individual farm or federal institution, the more financial, technical, human, social, institutional, and natural resources available, the better the chances for successful adaptation. Panellists also offered their insights on this aspect of the adaptation question.

Bruce Beattie, Alberta dairy farmer representing the Alberta Environmentally Sustainable Agriculture Council:
In terms of climate change, larger farms have some advantage for big tech solutions. For investing in some technological aspects (e.g., new equipment for conservation tillage), they can get access to financial capital and afford the operating cost more easily than a smaller mixed farming operation. Big is not always bad.

However, for some adaptive responses (such as the tendency to think you can rely on irrigation) well that is mistaken – we are pretty well maxed out in terms of water resources in southern Alberta.

Capacity is an interesting topic – lots of heat but not much light.

Nigel Weber, Alberta family farmer (mixed production) with the National Farmers Union:
Small and medium-sized family farms are adaptable, resilient, and committed to their location. Over the past generation, most have adapted to market forces by shifting production between a wide range of crops and livestock. Many have made large transitions – moving from grain farming to forage production or moving from dairy farming to beef production. As variability and uncertainty grows, the existence of adaptable, family farm agriculture becomes increasingly important.

Over the past century, Canadian farmers have eagerly adapted to a range of new technologies and practices. Anyone who has farmed or visited a farm regularly knows how much has changed from one decade to the next. Increasingly, however, financial pressures are preventing farmers from making certain changes on their farms that would be good, not only for their operations, but for the environment.

To adapt to any changes, farmers need prices high enough to give a farm family long-term stability and give it capital to invest. Fair and adequate commodity prices are *essential* if our farms are to adapt to climate change.

Don McCabe, Ontario cash crop producer with the Soil Conservation Council of Canada:

Canadian agriculture is the "deer in the headlights"; the farmer is the endangered species. We need to take this focus on adaptation one step further: as producers, we must change policy, with current conditions there is no longer any incentive for anyone to enter farming. Please note that increased yields do not mean more money in the farmers' pocket.

Bruce Burnett, Canadian Wheat Board:

Another point to make about drought in the growing season is that the exact opposite of drought can be just as devastating ... extremes are what we are worrying about.

Alan Stewart, Prairie Farm Rehabilitation Administration:

Governments, agribusiness, and farmer organizations are engaged in a continual process of searching out new crops, new varieties, new production methods, new controls for weeds and pests, new uses and end-products, and so on. This kind of work is an essential element of being prepared for climate change. As climate change unfolds, this on-going research will contribute to the myriad of small adjustments that collectively are the adaptation to gradual change. If climate change is more abrupt or unpredictably variable (as re-creations of Little Ice Age [Fagan] suggest it might be) then at the very least there will be some resource of experience and expertise.

To ensure we have this capacity, these institutions need to be targeted to include climate change in their vision and to assist in building the understanding of climate change impacts from their institutional/ member perspectives. A tangible example would be engaging irrigation districts in work on climate change and water resources. McLeman's work [in Chapter 15 of this volume] is a good reminder that there may be groups that are particularly vulnerable. It may be premature to suggest specific initiatives, but it seems reasonable to recognize the possibilities of many farm enterprises experiencing hardship and consider a range of tools from training and retraining to debt adjustment and re-investment that might be employed to assist change.

However, the role of "rural-town" or "rural-farm" communities in developing responses to climate change may be limited since the adaptation investments needed are by the individual production businesses rather than "communities." My experience has been that community-based action is effective when there is a specific, identifiable issue, and the issue is such that it clearly identifies the "community."

Brian Abrahamson, retired from the Prairie Farm Rehabilitation Administration:
Policy makers and industry can use both current and predicted future variability to make decisions with respect to policy and investment in infrastructure. Agricultural policy may be implemented in five- to ten-year increments with consideration of how current policy may influence future policy. For example, the likelihood of increased aridity in the Canadian Prairies suggests that soil conservation policies should be maintained or even enhanced as stresses increase on existing agricultural lands and new lands become vulnerable. Investment in infrastructure requires a longer time frame. For example, infrastructure such as a processing plant, a research facility, or an irrigation reservoir may be designed for longer periods of up to 30 years.

Nancy Lease, Quebec Ministry of Agriculture, Fisheries and Food:
In fact, policy makers may at times be chiefly interested by what the research cannot tell us, as policy decisions must often be taken even though several crucial pieces of information are not available. So as not to delay timely decision making, it is essential for research results, be they biophysical or socio-economic in nature, to honestly also present what cannot be captured within a reasonable level of confidence by scientific analysis. Despite the limitations inherent in this field (i.e., of climate change), research results to date have successfully served to alert us to the issue of climate change and to make us start thinking of climate in less static terms than we were accustomed to.

As a first suggestion for designing climate research to be of use to the agricultural policy community, I might emphasize the importance of careful planning and direction for climate vulnerability discussions and focus groups. Second, regardless of the care taken by researchers implicated in modelling studies to emphasize that their results are partial analysis based on scenarios and hypotheses, these studies are often cited as "predictions" of the future. This is yet another indication of the human desire to look into that crystal ball. The more that these studies present several results, for example, by using various models and emissions scenarios, the more we can hope that users will understand better the real nature of modelling research.

Elizabeth Atkinson, Natural Resources Canada:
   If the agri-food sector can't adapt in Canada (given all the capacity) then the rest of the world does not stand a chance ... One of the kinds of things we need to know and what I have heard today include ideas about capacity – what it is; who has it and where is it most beneficial and appropriately implemented ... capacity assessments could inform primarily through local grower engagement ... regional equity comes into play here.

Adaptation Processes

Information in Part 3 offers a view of how climate adaptation takes place in the agri-food sector and points out that, for producers, it is never in isolation from other risk management practices. For other dimensions of the sector (e.g., government), it is also the case that, in effect, climate change adaptation is implicated in many policies and programs that do not specify their climate change connection. Several panel members refer to issues that must be taken into account to understand the process of adapting to climate change more fully.

Alan Stewart, Prairie Farm Rehabilitation Administration:
   I think there are opportunities for this kind of community-based action as long as there are the definable issues and communities. Currently, stream health and water quality protection are being quite successfully addressed through community action in many areas. The need for, and extent of, community-based action on this issue may increase as climate change impacts surface water resources. Other issues such as pest mitigation or control may also lend themselves to a community-based approach, depending on what the specific local impacts are and how the issues define communities.
   At the local level, the roles of informal communities and individual business enterprises overlap. Local community-based approaches to addressing issues arising from climate impacts may be successful where the impacts clearly define the community. In other cases, it may well prove that adjustments made through individual business operations will be the main vehicle of change. At both the community and the individual business level developing new enterprises to take advantage of emerging opportunities can probably rely on existing investment mechanisms. However, climate change that limits current production systems and forces change to lower productivity systems will require

re-investment that many of the business owners may not be able to make. In this case, other mechanisms may be needed to make it possible for farm businesses to change, and to assist rural communities make adjustments.

Brian Abrahamson, retired from the Prairie Farm Rehabilitation Administration:
Producers adapt through farm management, governments and industry adapt through policy and infrastructure, and all use the results of research. It might be helpful to think of adaptation as a continuous process starting with those measures taken by producers to adapt to current climate variability and extending our thinking to what is required to adapt to an expected future scenario some 50 years from now. From previous workshops we have learned that producers are continually adapting to climate variability, markets, and farm policy. For producers current variability is the primary threat and the one to which they will focus their adaptation efforts.

Nancy Lease, Quebec Ministry of Agriculture, Fisheries and Food:
When climate change for the agricultural sector exists as an isolated or a driving factor for research, its ability to contribute to policy development is reduced. Agriculture, being a complex integration of biophysical, economic and social factors, is best studied using a systems approach. Partial analysis can lead not just to partial conclusions, but also to erroneous ones.

Therefore, climate change research may prove itself most relevant for policy development when it is presented as an integrated issue in the context of risk and uncertainty for farm management decisions. Studies that combine biophysical and socio-economic analysis and that aim to define resilient farming systems can be particularly valuable for policy development. Such information can contribute to the policy goal of assuring a stable food supply to consumers as well as a stable income to producers, using resources in a sustainable fashion, within a context of changing conditions, of which climate is one.

Elizabeth Atkinson, Natural Resources Canada:
Policy making is not a pretty sight; it can be an art form. Whatever issue is being addressed it is never in isolation. Policy responses are a mix of influences – institutional, cultural, economic, political, what have you. Understanding these elements and the context within which they work with the issue you are examining is really the art of good policy making.

What is interesting about what I have seen today is that the capacity assessment is very similar to that kind of analysis that is considering a multi-stressed system and striving to understand the elements that are working in it from all aspects and engaging the local community.

The material presented in this chapter suggests that climate change adaptation research, policy, and programs have made some progress but still have much to accomplish. Chapter 17 serves as a conclusion to this book by summarizing its findings and contributions and noting what remains to be addressed on the topic of climate change adaptation in the Canadian agri-food sector.

17
Climate Change Adaptation Research and Policy for Canadian Agriculture
Ellen Wall and Barry Smit

The adaptive capacity for dealing with current and future climatic and weather conditions is an important aspect of Canadian agricultural viability. Although impacts and opportunities from an altered climate affect different components of that sector (i.e., producers, industry, and government) in different ways, these components hold in common their appreciation for proactive adaptation and point to the importance of research for developing effective responses (C-CIARN Agriculture 2004). Previous chapters of this book contain recent research results about climate change adaptation and offer insights for building adaptive capacity for climate change in Canadian agriculture. The purpose of this chapter is to highlight the key messages from the book.

Although the material presented ranges over regions, farming systems, and research approaches, three overarching themes are evident: (1) climate and changing climatic conditions are important for agriculture; (2) producers deal with those changes as part of their broader risk management strategy; and (3) research and policy for climate change adaptation need more attention.

Climate Matters

Climate and weather conditions are key factors in agricultural production; these conditions appear to be changing and are projected to change more in the future. Extended droughts, variability, and increases in temperature are often noted as the conditions causing most concern for agricultural production, while longer growing seasons offer potential increases in yield and diversity of crops grown. Climate and weather impacts vary widely according to region and farming system.

Research regarding climate change impacts and their implications for adaptation to potential climate changes has focused primarily on crop production, with earlier analyses suggesting overall positive impacts (Chapter 2). The potential benefits to the industry from longer growing seasons and increased average temperatures may include increased crop yields as long as other conditions (e.g., adequate moisture levels and acceptable soil quality) exist and adaptation is practical.

Impact-based approaches for agriculture emphasize biophysical conditions. Samuel Gameda and colleagues' suggestion in Chapter 5 that under future climate projections, corn and soybeans would likely expand in eastern Canada might be different if the historical and continuing trend for reduced commodity prices (which might dissuade producers from growing those crops) is factored into the analysis. David Sauchyn (Chapter 6) considers other factors in his discussion of future climate impacts on crop production for the Canadian Prairies. He notes the crucial importance of adaptive strategies (at all levels) throughout Prairie farming history and suggests that future conditions will increase the need for effective adaptation. Sauchyn's impact-based research points out how future climate is important in terms of increased climatic variability and droughts relative to sensitive soil and water resources typical in Prairie agriculture. Denise Neilsen and colleagues (Chapter 7) focus on projected climate change impacts on water resources in the Okanagan Basin and show how implications for agricultural production are highly dependent on water availability. They conclude that existing infrastructure will be inadequate for supplying future demand, and use this projected impact as a starting point for engaging local users in discussion of possible adaptation strategies.

While context-based research acknowledges that climatic and weather conditions are important external factors for the agricultural sector, it treats them as one of many risks or opportunities that the sector must manage successfully in order to be sustainable. Ben Bradshaw (Chapter 8) suggests that producers' stress from increasingly depreciated commodity prices and other market-related factors necessarily limits their interest in the effects of future climate change. Henry Venema (Chapter 9) notes how prevailing agricultural farm practices (such as those limiting soil erosion) play an important role in determining how well the sector can respond to climatic and weather conditions. The negative impacts from future climate change will be lessened if farming systems are able to build resilience through suitable environmental practices and stronger economic resources. Christopher Bryant and colleagues (Chapter 11) describe how early research in Quebec focused solely on scenario-based impact assessments for projecting crop yield possibilities. To understand the implications of those impacts for the agricultural sector, however, researchers had to include a "human agency" dimension and investigate how producers might respond to the potential for improved conditions.

The importance of climatic and weather conditions for agricultural production is strongly supported by process-based research. While impact-based studies provide a macro and "top-down" level of analysis based on certain climate projections for the future, process-based inquiry proceeds from a "bottom-up," micro-level analysis documenting changes and impacts as producers experience them. In vulnerability studies (Suzanne Belliveau and colleagues in Chapter 12 and Susanna Reid and colleagues in Chapter 13), producers are specifically asked to describe the conditions (including climate and weather) that they find favourable and/or harmful for their farm operations and how they deal with the related challenges and opportunities. Documentation of producers' responses leaves no doubt that weather and climate are key elements. Using historical examples, Cynthia Neudoerffer and David Waltner-Toews (Chapter 14) and Robert McLeman (Chapter 15) also provide examples of the significant role agroclimatic conditions play in farming systems. Their respective work in current Manitoba farming communities and past Oklahoma communities points to the importance of extreme events. They conclude that a comprehensive examination of how agricultural communities respond to past agroclimatic conditions will improve such communities' capacity for responding to future climatic and weather conditions.

Producers Continue to Adapt

Producers manage climate and weather risks in a multiple-risk environment and continue to employ many strategies for adapting to changing conditions. Adaptation involves government, industry, and farm-level actions. Because producers adapt to climate change in conjunction with other business risk management strategies, it is important to use a "whole-farm" approach for understanding adaptation issues.

Generally speaking, impact-based research tends to treat climate factors in isolation and work from a "top-down" view. Michael Brklacich and colleagues (Chapter 2) note that initial research from an impact-based perspective generally ignored adaptive responses when projecting possible impacts for Canadian agriculture from future climate change. Possible increases and decreases in crop yield were determined and assessments made on the basis of the biophysical conditions favouring or hampering production (see Chapter 5). Although later impact-based research has attempted to include producers' adaptive responses in modelling, there are many assumptions about the kinds of options they might favour, so model results remain hypothetical.

Including adaptive responses in impact assessments can lead to a determination of "net" or "residual" impacts from climate change. Sauchyn (Chapter 6) suggests that this type of assessment could identify the need for increasing the current coping range for agricultural production if future climate change models are consistently projecting conditions that go beyond what the sector can handle now. Another trend with impact-based research is to use the results from future climate change scenarios as a tool for engaging producers and other agricultural industry representatives in discussion about possible adaptive strategies (see Chapter 7).

As the name of the context-based approach suggests, it places heavy emphasis on understanding and integrating multiple factors that affect the decision-making environments for agricultural producers. Bradshaw (Chapter 8) reviews past adaptive strategies for external factors other than climatic and weather conditions to get some sense of how the industry has responded to stress historically. The results challenge assumptions about what producers will do when faced with more desirable or more unfavourable climate and weather in the future. For instance, Bradshaw points out that sector trends towards specialization, concentration, and increased productivity could make it more difficult for producers to diversify their operations as a proactive adaptive response. As well, the continuing loss of farm income in combination with decreasing product prices further reduces the adaptive capacity of individual farmers. Venema (Chapter 9) builds on Bradshaw's argument, pointing out the challenges for building resilience in Prairie farming from external factors such as trade liberalization, resource depletion, and genetically engineered crop production.

The context for Harry Diaz and David Gauthier's discussion in Chapter 10 is primarily institutional. They suggest that without (environmentally sustainable) government support, many farming regions will have difficulty surviving under projected climate change conditions. Thus, climate change adaptation at the farm level will be greatly enhanced if adaptive capacity at the institutional level is also developed and maintained. They refer to networks of government and civil society organizations in the South Saskatchewan River Basin as examples of the type of institutional capacity that will enable agriculture to continue. Referring to a different Canadian region, Bryant and colleagues (Chapter 11) demonstrate the need to "ground understanding of farm adaptation and farm vulnerability [to climate change impacts] in the realities of farming and farmer decision making." Especially important for their work is understanding farmers' risk perception and risk management behaviour. Later research with Québec's crop insurance institutions targets areas and producers with persistent difficulty in handling current and recent weather challenges. Once these are known, efforts to enhance adaptive capacity for those conditions (for instance, through improved soil and water resource management) may reduce their reliance on crop insurance payments.

Process-based inquiry focuses on historical social and economic conditions within which farming operates, through intensive interactions with the stakeholders involved. Belliveau and colleagues (Chapter 12) use a whole-farm approach to discover how market pressures to meet consumer demand can prompt producers into planting varieties that are not necessarily suitable for existing or projected climatic and weather conditions. Likewise, Reid and colleagues (Chapter 13) show how farmers use a vast array of highly specific adaptations to manage climate and weather risks. They are employed according to the individual producers' access to resources from a personal level (financial capital, skill, management ability) to community level (support from social networks) to industry level (technology available) and institutional level (regulatory environment, program support).

Neudoerffer and Waltner-Toews (Chapter 14) also use in-depth knowledge of a Manitoba agricultural community to examine its past capacity for dealing with water resources as an indication of how it might fare under projected future conditions. They conclude that there will be several challenges for the community studied (loss of productive population, increase in farm concentration, rural community decline, and lack of government support) that will make responding to changing climatic and weather conditions even more difficult. McLeman (Chapter 15) provides a case study of farming communities directly affected by adverse climate and weather conditions during the 1930s Dust Bowl. Using archival and historical records as well as information from surviving community members, he shows that the ability to manage risk largely depended on access to different resources combined with a set of external conditions that restricted adaptive responses.

Research and Policy
Research and policy initiatives for adaptation to climate change are relatively undeveloped for a number of reasons, including the tendency for climate change to be equated only with reduction of greenhouse gas emissions, without acknowledgment of the need for understanding adaptation to altering conditions. The Canadian agricultural sector will benefit from policy-relevant research that examines producers' capacity to deal with climate and weather risks.

Each of the three research approaches featured in this book makes specific contributions to the field of study referred to as climate change adaptation for agriculture. Their roles in generating policy-relevant research are distinct, as are the gaps they point out in the knowledge base.

Impact-Based Approach
Research conducted from an impact-based perspective is designed to examine the biophysical and production effects of certain features of an agricultural system, such as changes in yield or the potential for regionally based production to respond to a stressor such as changes in temperature norms. Impact-based research provides big-picture assessments and overviews that are useful for communicating general trends and regional differences in production relative to changes in the variables included in climate change scenarios. Gameda and colleagues (Chapter 5) outline the potential for expanded crop production in central and eastern Canada, while Sauchyn (Chapter 6) cautions that the Prairie region remains increasingly vulnerable to climatic and weather conditions. Neilsen and colleagues (Chapter 7) indicate that under climate change projections, water demand from a number

of users will present serious challenges to irrigation-based agriculture in the Okanagan Basin of British Columbia.

The knowledge base for understanding climate change adaptation in Canadian agriculture will benefit from impact-based research that:

- includes agroclimatic variables that the agricultural sector identifies as important for production
- improves downscaling techniques to strengthen the connection between global models and local conditions
- builds up existing climate and weather data collection in order to assess the sensitivity of the sector to changes and extremes
- covers more agricultural regions in Canada
- incorporates scenarios for socio-economic trends
- includes analysis of adaptation processes and the non-climatic conditions that influence adaptation.

Context-Based Approach

The context-based approach for studies on climate change adaptation and agriculture goes beyond the influence of climate change and addresses the broad range of trends and conditions that affect producers' ability to adapt. Many of those factors are addressed in research on the political economy of Canadian agriculture, a field of study with a long and rich history.

Among the challenges for context-based research is where to draw boundaries for data gathering and analysis. What is *not* relevant for understanding the context for climate change adaptation in Canadian agriculture? Bradshaw (Chapter 8) and Venema (Chapter 9) address the influence from a vast range of external factors related to social, political, economic, and environmental conditions from local to international scales. Diaz and Gauthier (Chapter 10) provide very detailed assessments of institutional capacity to support their claim that it is a prerequisite for individual and community-level adaptation. Bryant and colleagues (Chapter 11) suggest that research into climate change adaptation and Québec agriculture has become increasingly holistic. For their current research, the context for understanding individual risk management includes comprehensive data sets on crop production, financial losses, and environmental conditions for Québec and its regions over time.

The knowledge base for understanding climate change adaptation in Canadian agriculture will benefit from context-based research that:

- highlights the ways in which climate change fits in with the multiple stresses
- examines producers' risk perceptions and management in more detail

- identifies some general insights and lessons for adaptation beyond individual case studies
- connects the context insights to policy and management opportunities for adaptation.

Process-Based Approach

Climate change and adaptation studies for Canadian agriculture that follow the process-based approach identify a specific system of interest and then (as with the context-based approach) consider a wide range of influences on that system. Questions about what *not* to include are therefore also relevant for process-based research. One advantage that this perspective may have in establishing such limits is its reliance on the stakeholders involved to determine what the key factors are. Belliveau and colleagues (Chapter 12) and Reid and colleagues (Chapter 13) employed this technique in their studies by not introducing the topic of climatic and weather conditions as stressors when starting discussion with producers in their samples. Instead, those interviewed inevitably raised the topic of how weather and climate affect production. In this way, informants also revealed how climate and weather issues were related to or integrated with other production concerns, thereby increasing researcher understanding of how producers view their operations, and where and how climate and weather fit in.

Compared with impact-based inquiry, very little research in Canadian agriculture has followed a process-based approach. Thus, one major gap in this type of study relates to how little of it has been done. Also, given the idiosyncratic nature of the data gathered, it is not always possible to extrapolate from the individual cases to obtain a more general picture.

The knowledge base for understanding climate change adaptation in Canadian agriculture will benefit from process-based research that:

- covers the major agricultural regions and types of commodity production in Canada
- recognizes the broader forces that drive or constrain local exposures and adaptive strategies
- applies knowledge of current stimuli and processes to future conditions
- connects with decision-making (at the farm, business, or government levels) structures and institutions in order that climate change adaptation can be better considered.

The Canadian agricultural sector is expected to realize some benefits from climate change, notably from longer growing seasons and more heat available for plant growth. It is apparent, however, that climate change will likely exacerbate certain climate-related risks, particularly those associated with moisture and extremes, that the industry already has to deal with. At the farm level, adaptation is essentially a management process, something producers do on a routine basis. Climate change modifies some of the hazards and introduces some new risks and opportunities.

The research outlined in this book shows that the knowledge base for climate change impacts, vulnerabilities, and adaptations is still evolving. Notwithstanding developments in stakeholder engagement, most of the research is not well integrated into decision making by farming communities, businesses, and governments. Partly because of this limited integration, climate change adaptation research has yet to have a major influence on agricultural policy and programs in Canada, at the federal or provincial levels. As Canadian producers and the agri-food sector experience the effects of climate change more widely, the need for knowledge of adaptation initiatives will increase. Information presented in this book provides a strong basis for producers, the agri-food industry, and policy makers to build their capacity for dealing with the risks and opportunities associated with farming in a changing climate.

References

AAFC (Agriculture and Agri-Food Canada). 2002. Irrigation system operates on a "need to drink" basis. *Agvance* 9 (3): 4.
–. 2003. Innovation in agriculture: Finding the better way. *Canadian Farm Manager* (April/May).
–. 2005. *Climate change related programs*. http://www.agr.gc.ca/cc_programs_e.php.
Adams, R., B. Hurd, S. Lenhart, and N. Leary. 1998. Effects of global climate change on agriculture: An interpretive review. *Climate Research* 11: 19-30.
Adams, R., B. McCarl, D. Dudek, and J. Glyer. 1988. Implications of global climate change for western agriculture. *Western Journal of Agricultural Economics* 13: 348-56.
Adams, R., B. McCarl, and L. Mearns. 2003. The effects of spatial scale of climate scenarios on economic assessments: An example from US agriculture. *Climatic Change* 60: 131-48.
Adger, W. 2003. Social aspects of adaptive capacity. In *Climate change, adaptive capacity and development*, ed. J. Smith, R. Klein, and S. Huq, 29-49. London: Imperial College Press.
Adger, W.N. 1999. Social vulnerability to climate change and extremes in coastal Vietnam. *World Development* 27: 249-69.
–. 2001. Scales of governance and environmental justice for adaptation and mitigation of climate change. *Journal of International Development* 13: 921-31.
Adger, W.N., N. Brooks, G. Bentham, M. Agnew, and S. Eriksen. 2004. *New indicators of vulnerability and adaptive capacity*. Tyndall Centre for Climate Change Research technical report no. 7. Norwich, UK: Tyndall Centre.
Adger, W.N., and P.M. Kelly. 1999. Social vulnerability to climate change and the architecture of entitlements. *Mitigation and Adaptation Strategies for Global Change* 4 (3/4): 253-66.
Adger, W.N., P.M. Kelly, and N.H. Ninh, eds. 2001. *Living with environmental change: Social resilience, adaptation and vulnerability in Vietnam*. London: Routledge.
Agrodev Canada Inc. 1994. *Water supply and management issues affecting the BC tree fruit industry. Final report*. Summerland, BC: Okanagan Valley Tree Fruit Authority.
Agronomics Interpretations Working Group. 1995. Land suitability rating system for agricultural crops: 1. Spring-seeded small grains. Ed. W.W. Pettapiece. Technical bulletin 1995-6E. Ottawa: Centre for Land and Biological Resources Research, Agriculture and Agri-Food Canada. Online: http://sis.agr.gc.ca/cansis/publications/manuals/lsrs.html.
Ahmad, Q., et al. 2001. Summary for policy makers. In *Climate change 2001: Impacts, adaptation and vulnerability. Intergovernmental Panel on Climate Change, Working Group II, Third Assessment Report*, ed. J. McCarthy, O. Canziani, N. Leary, D. Dokken, and K. White, secs. 1, 2, 5. Cambridge, UK: Cambridge University Press.
Albritton, D.L., and L.G. Meira Filho. 2001. Technical summary. In *Climate change 2001: The scientific basis*, ed. J. Houghton, Y. Ding, J. Griggs, M. Noguer, P. van der Linden, and D. Xiaosu, 1-21. Cambridge, UK: Cambridge University Press. [see Beade et al. 2001 for comparison]

Alcorn, J., and V. Toledo. 2000. Resilient resource management in Mexico's forest ecosystems. In *Linking social and ecological systems: Management practices and social mechanisms for building resilience*, ed. F. Berkes and C. Folke, 216-49. Cambridge, UK: Cambridge University Press.

Alfaro, S. 2004. *Institutions and sustainability. An analytical report*. Working paper. Regina: Institutional Adaptation to Climate Change Project.

Allen R.G., L.S. Pereira, D. Raes, and M. Smith. 1998. *Crop evapotranspiration – guidelines for computing crop water requirements*. FAO Irrigation and Drainage paper no. 56. Rome: United Nations Food and Agriculture Organization.

Alwang, J., P.B. Siegel, and S.L. Jorgensen. 2002. *Vulnerability as viewed from different disciplines*. Washington, DC: World Bank.

André, P., and C.R. Bryant. 2001. *Évaluation environnementale des stratégies d'investissement des producteurs agricoles de la région de Montréal en regard des changements climatiques*. Research report submitted to the Climate Change Action Fund, Natural Resources Canada, May 2001.

André, P., B. Singh, C.R. Bryant, J.P. Thouez, and M. El-Mayaar. 1996. Impact of climate change on Quebec agriculture: The integration of social and physical dimensions. In *IAIA '96 Improving environmental assessment effectiveness: Research, practice and training. Conference proceedings: Volume 2*. Fargo, ND: International Association for Impact Assessment.

Andresen, J., G. Alagarswamy, D. Stead, H. Cheng, and B. Sea. 2000. *Preparing for a changing climate: The potential consequences of climate variability and change in the Great Lakes Region (Agriculture)*. Great Lakes Regional Assessment, http://www.geo.msu.edu/glra/assessment/assessment.html.

Antle, J.M., S.M. Capalbo, and J. Hewitt. 2004. Adaptation, spatial heterogeneity, and the vulnerability of agricultural systems to climate change and CO_2 fertilization: An integrated assessment approach. *Climatic Change* 64 (3): 289-315.

Archambault, D.J., L. Xiaomei, D. Robinson, J.T. O'Donovan, and K.K. Klein. 2001. *The effects of elevated CO_2 and temperature on herbicide efficacy and weed/crop competition*. Final report prepared for PARC (Prairie Adaptation Research Collaborative). University of Regina. http://www.parc.ca/pdf/research_publications/agriculture2.pdf.

Arthur, L., and F. Abizadeh. 1988. Potential effects of climate change on agriculture in the Prairie region of Canada. *Western Journal of Agricultural Economics* 13: 216-24.

Asselstine, A. 2003. Canada's crop insurance and disaster assistance programs. Presented at the Australian Agricultural and Resource Economics Society pre-conference workshop, "Managing Climate Risks in Agriculture," Ferndale, Western Australia, February.

Associated Engineering. 1997. *Water system master plan*. Unpublished report to the District of Summerland, BC.

Association of British Columbia Grape Growers. 1984. *Atlas of suitable grape growing locations in the Okanagan and Similkameen valleys of British Columbia*. Kelowna: Association of British Columbia Grape Growers.

Aubin, P., G. Auger, and C. Perreault. 2003. *Climate change and greenhouse gas awareness study report*. Agriculture and Agri-Food Canada, http://www.cciarn.uoguelph.ca/documents/clima_e.pdf.

Baker, B.B., J.D. Hanson, R.M. Bourdon, and J.B. Eckert. 1993. The potential effects of climate change on ecosystem processes and cattle production on US rangelands. *Climatic Change* 25 (2): 97-117.

Barnett, T., R. Malone, W. Pennell, D. Stammer, B. Semtner, and W. Washington. 2004. The effects of climate change on water resources in the west: Introduction and overview. *Climatic Change* 62: 1-11.

Bauder, H. 2003. "Brain abuse," or the devaluation of immigrant labour in Canada. *Antipode* 35: 699-717.

BCMAFF (British Columbia Ministry of Agriculture, Food and Fisheries). 2002. *Annual BC horticultural statistics*. http://www.agf.gov.bc.ca/stats/wholehort.pdf.

Beade, A., E. Ahlonsou, Y. Ding, and D. Schimel. 2001. The climate system: An overview. In *Climate change 2001: The scientific basis*, ed. J. Houghton, Y. Ding, J. Griggs, M. Noguer, P. van der Linden, and D. Xiaosu, 85-98. Cambridge, UK: Cambridge University Press.

Behboudian, M.H., J. Dixon, and K. Pothamshetty. 1998. Plant and fruit responses of lysimeter grown "Braeburn" apple to deficit irrigation. *Journal of Horticultural Science and Biotechnology* 73: 781-85.

Bélanger, G., P. Rochette, Y. Castonguay, A. Bootsma, D. Mongrain, and D.A.J. Ryan. 2001. *Impact of climate change on risk of winter damage to agricultural perennial plants.* Climate Change Action Fund Project A084 final report. Sainte-Foy, QC: Soils and Crops Research and Development Centre, Agriculture and Agri-Food Canada.

–. 2002. Climate change and winter survival of perennial forage crops in eastern Canada. *Agronomy Journal* 94 (5): 1120-30.

Belliveau, S., Smit, B., and B. Bradshaw. 2006. Multiple exposures and dynamic vulnerability: Evidence from the grape industry in the Okanagan Valley, Canada. *Global Environmental Change* 16: 364-78.

Berkes, F., J. Colding, and C. Folke. 2003. *Navigating social-ecological systems: Building resilience for complexity and change.* Cambridge, UK: Cambridge University Press.

Bhatti, J., R. Lal, M. Apps, and M. Price, eds. 2006. *Climate change and managed ecosystems.* New York: Taylor and Francis Group.

Blaikie, P.M., and H.C. Brookfield. 1987. *Land degradation and society.* London: Methuen.

Blaikie P., T. Cannon, I. Davis, and B. Wisner. 1994. *At risk: Natural hazards, people's vulnerability, and disaster.* London: Routledge.

Boer, G.J., G. Flato, and D. Ramsden. 2000. A transient climate change simulation with greenhouse gas and aerosol forcing: Projected climate to the twenty-first century. *Climate Dynamics.* 16: 427-50.

Boland, G., M. Melzer, V. Higgins, A. Hopkin, and A. Nassuth. 2003. Climate change and plant disease in Ontario. In *A synopsis of the known and potential diseases and parasites associated with climate change,* ed. S. Griefenhagen and T. Noland, 7-89. Forest research information paper no. 154. Sault Ste. Marie: Ontario Ministry of Natural Resources.

Bollig, M., and A. Schulte. 1999. Environmental change and pastoral perceptions: Degradation and indigenous knowledge in two African pastoral communities. *Human Ecology* 27 (3): 493-514.

Bollman, R., L. Whitener, and F. Tung. 1995. Trends and patterns of agricultural structural change: A Canada-US comparison. *Canadian Journal of Agricultural Economics* (Special Issue): 15-28

Bond, J.H., R. McKinley, and E.H. Banks. 1940. Testimony of J.H. Bond, Assistant Director; R. McKinley, Farm Placement Supervisor; and E.H. Banks, Farm Placement Supervisor, Texas State Employment Service, Austin. In *Select Committee to Investigate the Interstate Migration of Destitute Citizens,* 1812-32. US House of Representatives, 67th Congress, Oklahoma City hearings.

Bootsma, A. 2002. A summary of some results of research on the potential impacts of climate change on agriculture in eastern Canada. Presented at the "Climate and Agriculture in the Great Lakes Region: The Potential Impacts and What We Can Do" workshop at Michigan State University, East Lansing, 22 March. Online: http://www.geo.msu.edu/glra/workshop/03agriworkshp/agenda.htm.

Bootsma, A., J. Boisvert, and J. Dumanski. 1994. Climate-based estimates of potential forage yields in Canada using a crop growth model. *Agriculture and Forest Meteorology* 67: 151-72.

Bootsma, A., S. Gameda, and D.W. McKenney. 2005a. Impacts of potential climate change on selected agroclimatic indices in Atlantic Canada. *Canadian Journal of Soil Science* 85: 329-43.

–. 2005b. Potential impacts of climate change on corn, soybeans and barley yields in Atlantic Canada. *Canadian Journal of Soil Science* 85: 345-57.

Bootsma, A., S. Gameda, D. McKenney, P. Schut, H. Hayhoe, R. de Jong, and E. Huffman. 2001. *Adaptation of agricultural production to climate change in Atlantic Canada.* Final report submitted to the Climate Change Action Fund, Natural Resources Canada.

Bootsma, A., R. Gordon, G. Read, and W.G. Richards. 1992. *Heat units for corn in the Maritime provinces.* Publication 92-1. Nova Scotia: Atlantic Advisory Committee on Agrometeorology.

Bootsma, A., G. Tremblay, and P. Filion. 1999. *Risk analyses of heat units available for corn and soybean production in Quebec.* Technical Bulletin, ECORC Contrib. No. 991396. Ottawa: Eastern Cereal and Oilseed Research Centre, Agriculture and Agri-Food Canada. Online: http://res2.agr.gc.ca/ecorc/clim2/index_e.htm.

Bourdieu, P. 1986. The forms of capital. In *Handbook of theory and research for the sociology of education,* ed. J.G. Richardson, 241-58. New York: Greenwood Press.

Bowen, P., C. Bogdanoff, B. Estergaard, S. Marsh, K. Usher, S. Smith, and G. Frank. 2005. Geology and wine – the use of geographic information system technology to assess viticultural performance in the Okanagan and Similkameen valleys. *Geoscience Canada* 32: 161-76. Bowler, I. 1985 Agriculture under the CAP: A geography. Manchester, UK: Manchester University Press.

Bradshaw, B., H. Dolan, and B. Smit. 2004. Farm-level adaptation to climatic variability and change: Crop diversification in the Canadian Prairies. *Climatic Change* 67: 119-41.

Bradshaw, B., and B. Smit. 1997. Subsidy removal and agroecosystem health. *Agriculture, Ecosystems and Environment* 64 (3): 245-60.

British Columbia Ministry of Sustainable Resource Management. 2001. *Terrestrial ecosystem mapping for the Okanagan Valley.* NTS map sheet 82E. Digital data available from Land Information BC, Victoria.

Brklacich, M., C.R. Bryant, B. Veenhof, and A. Beauchesne. 1997a. *Implications of global climate change for Canadian agriculture: A review and appraisal of research from 1984 to 1997: Volume 1: Synthesis and research needs; Volume 2: Research and report summaries.* Report submitted to the Environmental Adaptation Research Group, Atmospheric Environment Service, Environment Canada, July 1997. Published as: Implications of global climate change for Canadian agriculture: A review and appraisal of research from 1984 to 1997 – Volume 1: Synthesis and research needs. In *Canada country study: Climate impacts and adaptation. Volume 7, National Sectoral Volume,* ed. G. Koshida and W. Avis, 219-56. Downsview, ON: Environment Canada.

–. 2000. Agricultural adaptation to climate change: A comparative assessment of two types of farming in central Canada. In *Agricultural and environmental sustainability in the new countryside,* ed. H. Millward, K. Beesley, B. Ilbery, and L. Harrington. Winnipeg: Hignell Printing.

Brklacich, M., D. McNabb, C. Bryant, J. Dumanski. 1997b. Adaptability of agricultural systems to global climate change: A Renfrew County, Ontario, Canada pilot study. In *Agricultural restructuring and sustainability: A geographical perspective,* ed. B. Ilbery, Q. Chiotti, and T. Rickard, 212-56. Wallingford, Oxon, UK: CAB International.

Brklacich, M., and B. Smit. 1992. Implications of changes in climatic averages and variability on food production opportunities in Ontario, Canada. *Climatic Change* 20:1-21.

Brklacich, M., and R.B. Stewart. 1995. Impacts of climate change on wheat yields in the Canadian Prairies. In *Climate change and agriculture: Analysis of potential international impacts,* ed. C. Rosenzweig, 147-62. Special publication 59. Madison, WI: American Society of Agronomy.

Brown, D.M., and A. Bootsma. 1993. *Crop heat units for corn and other warm-season crops in Ontario.* Ontario Ministry of Agriculture and Food factsheet 93-119, Agdex 111/31. http://www.omafra.gov.on.ca/english/crops/facts/93-119.htm.

Bryant, C.R., and P. André. 2003. Adaptation and sustainable development of the rural community. In *Proceedings of the annual colloquium of the IGU Commission on the Sustainability of Rural Systems,* ed. L. Laurens and C.R. Bryant, 449-60. Montpellier: Université de Montpellier, and Montreal: IGU Commission on the Sustainability of Rural Systems.

Bryant, C.R., P. André, D. Provençal, B. Singh, J.P. Thouez, and M. El Mayaar. 1997. L'adaptation agricole aux changements climatiques: Le cas du Québec. *Le Climat* 14 (2): 81-97.

Bryant, C.R., P. André, J.-P. Thouez, B. Singh, S. Frej, D. Granjon, J.P. Brassard, and G. Beaulac. 2004. Agricultural adaptation to climate change: The incidental consequences of managing risk. In *The structure and dynamics of rural territories: Geographical perspectives,* ed. D. Ramsey and C.R. Bryant, 260-71. Brandon, MB: Rural Development Institute, Brandon University.

Bryant, C.R., and T.R.R. Johnston. 1992. *Agriculture in the city's countryside*. London, ON: Pinter Publishers.

Bryant, C.R., B. Smit, M. Brklacich, T. Johnston, J. Smithers, Q. Chiotti, and B. Singh. 2000. Adaptation in Canadian agriculture to climatic variability and change. *Climatic Change* 45: 181-201.

Burton, I., S. Huq, B. Lim, O. Pilifosova, and E.L. Schipper. 2002. From impacts assessment to adaptation priorities: The shaping of adaptation policy. *Climate Policy* 2: 145-59.

Buttel, F. 1997. Social institutions and environmental change. In *The international handbook of environmental sociology*, ed. M. Redclift and G. Woodgate,40-54. Cheltenham, UK: Edward Elgar.

Campbell, C., R. Zentner, S. Gameda, B. Blomert, and D. Wall. 2002. Production of annual crops on the Canadian Prairies: Trends during 1976-1998. *Canadian Journal of Soil Science* 82: 45-57.

Canada Land Inventory. 1966. *Land suitability of agriculture, NTS Map sheet 82E*. Ottawa: Environment Canada.

–. 1975. *Soil capability for agriculture, 1:1,000,000 map series*. Ottawa: Soil Research Institute, Research Branch, Agriculture Canada; Lands Directorate, Environment Canada.

Canadian Council of Ecological Areas. 2004. Ecozones of Canada. Canadian Council of Ecological Areas, Environment Canada, http://www.ccea.org/ecozones/index.html.

Canadian Institute for Climate Studies. 2002. Website for the Canadian Climate Impacts Scenarios Project (CCIS), http://www.cics.uvic.ca/scenarios/index.cgi.

Canadian Wheat Board. 2003. Payments and statistical tables, http://www.cwb.ca (accessed 14 February 2003).

Carpenter, S., B. Walker, J.M. Anderies, and N. Abel. 2001. From metaphor to measurement: Resilience of what to what? *Ecosystems* 4: 765-81.

Carter, T.R., M.L. Parry, H. Harasawa, and S. Nishioka. 1994. *IPCC technical guidelines for assessing climate change impacts and adaptations*. London: University College and Centre for Global Environmental Research.

CCIAD (Climate Change Impacts and Adaptation Directorate). 2002. *Climate change impacts and adaptation: A Canadian perspective – agriculture*. Ottawa: Natural Resources Canada.

C-CIARN Agriculture (Canadian Climate Impacts and Adaptation Research Network for Agriculture). 2002. *Report from producer survey at Canada's outdoor farmshow, Woodstock, Ontario*. September. University of Guelph, Guelph, ON: C-CIARN Agriculture.

–. 2003. *Meeting the challenges from climate change*. Summary report from C-CIARN Agriculture round table, Grainworld, Winnipeg, 25 February. University of Guelph, Guelph, ON: C-CIARN Agriculture.

–. 2004. *Climate change adaptation – a producer perspective on policy and programs*. Report on C-CIARN Agriculture meeting, Gatineau, Québec, 25 February. University of Guelph, Guelph, ON: C-CIARN Agriculture.

Chakraborty, S., A.V. Tiedemann, and P.S. Teng. 2000. Climate change: Potential impact on plant diseases. *Environmental Pollution* 108 (3): 317-26.

Chance, K. 2003. Managing climate risk in agriculture. *Connections* 5 (3): 20-22.

Chapman, L.J., and D.M. Brown. 1978. *The climates of Canada for agriculture*. Canada Land Inventory report no. 3. Ottawa: Lands Directorate, Environment Canada.

Charron, D., D. Waltner-Toews, A. Maarouf, and M. Stalker. 2003. A synopsis of the known and potential diseases and parasites of humans and animals associated with climate change in Ontario. In *A synopsis of the known and potential diseases and parasites associated with climate change*, ed. S. Griefenhagen and T. Nolan, 7-89. Forest research information paper no. 154. Toronto: Ontario Ministry of Natural Resources.

Chetner, S., and the Agroclimatic Atlas Working Group. 2003. *Agroclimatic atlas of Alberta, 1971 to 2000*. Agdex 071-1. Edmonton: Alberta Agriculture, Food and Rural Development.

Chiotti, Q. 1998. An assessment of the regional impacts and opportunities from climate change in Canada. *Canadian Geographer* 42 (4): 380-93.

Chiotti, Q., and T. Johnston. 1995. Extending the boundaries of climate change research: A discussion of agriculture. *Journal of Rural Studies* 11 (3): 335-50.

Chiotti, Q., T.R. Johnston, B. Smit, and B. Ebel. 1997. Agricultural response to climate change: A preliminary investigation of farm-level adaptation in southern Alberta. In *Agricultural restructuring and sustainability: A geographical perspective*, ed. B. Ilbery, Q. Chiotti, and T. Rickard, 167-83. Wallingford, Oxon, UK: CAB International.

Chow, R., H. Rees, and J.L. Daigle. 1999. Effectiveness of terraces/grassed waterway systems for soil and water conservation: A field evaluation. *Journal of Soil and Water Conservation* 54 (3): 577-83.

Chow, R., H. Rees, and J. Monteith. 2000. Seasonal distribution of runoff and soil loss under four tillage treatments in the upper St. John River valley, New Brunswick, Canada. *Canadian Journal of Soil Science* 80: 649-60.

CIMMYT (International Wheat and Maize Improvement Center). 2004. In quest for drought-tolerant varieties, CIMMYT sows first transgenic wheat field trials in Mexico. CIMMYT, http://www.cimmyt.org/english/webp/support/news/dreb.htm.

Clark, W. 1985. Scales of climate change. *Climatic Change* 7: 5-27.

Cline, W. 1992. *The economics of global warming*. Washington, DC: Institute for International Economics.

Cloutis, E., A. Kirch, G. Wiseman, J. Golby, and D. Carter. 2001. *Socio-economic vulnerability of Prairie communities to climate change*. Prairie Adaptation Research Collaborative, http://www.parc.ca/publications.htm.

Coakley, S.M., H. Scherm, and S. Chakraborty. 1999. Climate change and plant disease management. *Annual Review of Phytopathology* 37: 399-426.

Cochrane, W. 1958. *Farm prices: Myth and reality*. Minneapolis: University of Minnesota Press.

Cohen, R.D.H., C.D. Sykes, E.E. Wheaton, and J.P. Stevens. 2002. *Evaluation of the effects of climate change on forage and livestock production and assessment of adaptation strategies on the Canadian Prairies*. Report to the Prairie Adaptation Research Collaborative (PARC). http://www.parc.ca/pdf/research_publications/renamed/PARC-36.pdf.

Cohen, S., D. Neilsen, and R. Welbourn, eds. 2004. *Expanding the dialogue on climate change and water management in the Okanagan Basin, British Columbia*. Project A463/433, submitted to Adaptation Liaison Office, Climate Change Action Fund, Natural Resources Canada. http://www.ires.ubc.ca.

Cohen, S., D. Neilsen, S. Smith, T. Neale, B. Taylor, M. Barton, W. Merritt, Y. Alila, P. Shepherd, R. McNeill, J. Tansey, J. Carmichael, and S. Langsdale. 2006. Learning with local help: Expanding the dialogue on climate change and water management in the Okanagan Region, British Columbia, Canada. *Climatic Change* 75 (3): 331-58.

Cohen, S.J., K.A. Miller, A.F. Hamlet, and W. Avis. 2000. Climate change and resource management in the Columbia River Basin. *Water International* 25: 253-72.

Coleman, J.S. 1988. Social capital in the creation of human capital. *American Journal of Sociology* 94: S95-S120.

Coleman, W.J., and H.A. Hockley. 1940. *Legal aspects of landlord-tenant relationships in Oklahoma*. Report no. B-241. Stillwater: Oklahoma Agricultural Experiment Station.

Collins, T. 1936. Weekly administrative reports. Arvin Federal Migrant Labor Camp, Farm Security Administration, San Francisco.

Crutzen, P. 2002. Geology of mankind. *Nature* 415: 23.

CSIDC (Canada-Saskatchewan Irrigation Diversification Centre). 2000. Overview publication. Outlook, SK: CSIDC.

Cummings, H., and Associates. 2000. *The economic impacts of agriculture on the economy of Perth County*. http://www.ofa.on.ca/perth/ (accessed October 2001).

Daly, C., W.P. Gibson, G.H. Taylor, G.L. Johnson, and P. Pasteris. 2002. A knowledge-based approach to the statistical mapping of climate. *Climate Research* 22: 99-113.

Daly, C., R.P. Neilson, and D.L. Phillips. 1994. A statistical-topographic model for mapping climatological precipitation over mountainous terrain. *Journal of Applied Meteorology* 33: 140-58.

Davis, M. 2002. *Late Victorian holocausts: El Niño famines and the making of the Third World*. New York: Verso.

de Loë, R. 2005. *Assessment of agricultural water use/demand across Canada*. Final report for Sustainable Water Use Branch Water Policy and Coordination Directorate, Environment Canada. Gatineau, QC.

de Loë, R., and R. Kreutzwiser. 2000. Climate variability, climate change and water resource management in the Great Lakes. *Climatic Change* 45: 163-72.

de Loë, R.C., and L.C. Moraru. 2004. *Water use and sustainability issues in the Canadian agriculture sector.* Final report prepared for Sustainable Water Use Branch Water Policy and Coordination Directorate, Environment Canada, Gatineau, Québec.

de Jong, R., A. Bootsma, T. Huffman, and G. Roloff. 1999. *Crop yield variability under climate change and adaptive crop management scenarios*. Final project report submitted to the Climate Change Action Fund. Ottawa: Natural Resources Canada.

de Vries, J. 1985. Historical analysis of climate-society interaction. In *Climate impact assessment,* ed. R.W. Kates, J.H. Ausubel, and M. Berberian, 273-91. New York: Wiley.

Delcourt, G., and G.C. van Kooten. 1995. How resilient is grain production to climate change? Sustainable agriculture in a dryland cropping region of western Canada. *Journal of Sustainable Agriculture* 5 (3): 37-57.

Demuth, M.N., and A. Pietroniro. 2001. *The impact of climate change on the glaciers of the Canadian Rocky Mountain eastern slopes and implications for water resource–related adaptation in the Canadian Prairies.* Report to the Prairie Adaptation Research Collaborative on PARC project PARC-55.

DesJarlais, C., A. Bourque, R. Decoste, C. Demers, P. Deschamps, and K. Lam, eds. 2004. *Adapting to climate change.* Report from Ouranos. http://www.ouranos.ca/cc/climang5. pdf.

Dessai, S., and M. Hulme. 2004. Does climate adaptation policy need probabilities? *Climate Policy* 4: 107-28.

Diaz, P., and M. Nelson. 2005. The changing Prairie social landscape of Saskatchewan: The social capital of rural communities. *Prairie Forum* 30 (1).

Dolan, A., B. Smit, M. Skinner, B. Bradshaw, and C. Bryant. 2001. *Adaptation to climate change in agriculture: Evaluation of options.* Department of Geography occasional paper no. 26, University of Guelph.

Doorenbos, J., and W.O. Pruitt. 1977. *Guidelines for predicting crop water requirements.* FAO Drainage and Irrigation paper no. 24. Rome: United Nations Food and Agriculture Organization.

Downing, T.E. 2001. *Climate change vulnerability: Linking impacts and adaptation.* Report to the Governing Council of the United Nations Environment Programme. Oxford: Environmental Change Institute.

Driver, D. 2004. The farm crisis hits the pews. *United Church Observer,* June. Online: http://www.ucobserver.org/archives/jun04_nation.htm (accessed 23 November 2005).

Duncan, O.D. 1935. *Population trends in Oklahoma*. Report no. B-224. Stillwater: Oklahoma Agricultural Experiment Station.

–. 1943. *Recent population trends in Oklahoma.* Report no. B-269. Stillwater: Oklahoma Agricultural Experiment Station.

Eakin, H. 2000. Smallholder maize production and climatic risk: A case study from Mexico. *Climatic Change* 45: 19-36.

Easterling, W. 1996. Adapting North American agriculture to climate change in review. *Agricultural and Forest Meteorology* 80: 1-54.

Easterling, W.E., P.R. Crosson, N.J. Rosenberg, M.S. McKenney, L.A. Katz, and K.M. Lemon. 1993. Agricultural impacts of and response to climate change in the Missouri-Iowa-Nebraska-Kansas (MINK) region. *Climatic Change* 24 (1/2): 23-62.

Easterling, W.E., C.J. Hays, M. McKenney-Easterling, and J.R. Brandle. 1997. Modelling the effect of shelterbelts on maize productivity under climate change: An application of the EPIC model. *Agriculture, Ecosystems and Environment* 61 (2-3): 163-76.

Easterling, W.E., B. Hurd, and J. Smith. 2004. *Coping with global climate change: The role of adaptation in the United States.* Prepared for the Pew Center on Global Climate Change. http://www.pewclimate.org/docUploads/Adaptation%2Epdf.

Easterling, W.E., N.J. Rosenberg, K. Lemon, and M.S. McKenney. 1992a. Simulations of crop responses to climate change: Effects with present technology and currently available adjustments (the "smart farmer" scenario). *Agricultural and Forest Meteorology* 59: 75-102.

Easterling, W.E., N.J. Rosenberg, M.S. McKenney, C.A. Jones, P.T. Dyke, and J.R. Williams. 1992b. Preparing the erosion productivity impact calculator (EPIC) model to simulate crop response to climate change and the direct effects of CO "SUB 2." *Agricultural and Forest Meteorology* 59: 17-34.

Edey, S.N. 1977. *Growing degree-days and crop production in Canada*. Publication no. 1635. Ottawa: Agriculture Canada.

El Mayaar, M., B. Singh, P. André, C.R. Bryant and J.-P. Thouez. 1997. The effects of climate change and CO_2 fertilisation on agriculture in Québec. *Agricultural and Forest Meteorology* 85: 193-208.

Ellsworth, J.O., and F. Elliot. 1929. *Types-of-farming in Oklahoma*. Report no. B-181. Stillwater: Oklahoma Agricultural Experiment Station.

Environment Canada. 1994. *Canadian monthly climate data and 1961-1990 normals*. CD-ROM, ver. 3.0E. Downsview, ON: Atmospheric Environment Service, Environment Canada.

EU (European Union) Commission. 2001. *Risk management tools for EU agriculture*. Agriculture Directorate-General, http://europa.eu.int/comm/agriculture/publi/insurance/text_en.pdf.

Evenson, R.E. 1999. Global and local implications of biotechnology and climate change for future food supplies. *Proceedings of the National Academy of Sciences* 96 (11): 5921-28.

Expert Committee on Soil Survey. 1991. *Soil water investigation methods manual*. Comp. R.G. Eilers, chairman of Soil Water Interest Group, Land Resource Research Centre. Ottawa: Agriculture Canada.

Fairchild, G. 2004. Agriculture in Atlantic Canada. Presented at "Climate Change Adaptation – a Producer Perspective on Policy and Programs," C-CIARN Agriculture meeting, Gatineau Québec, 25 February.

FAO (Food and Agriculture Organization of the United Nations). 2003. *World agriculture towards 2015/2030*. London: Earthscan.

Feenstra, J., I. Burton, J. Smith, and R. Tol. 1998. *Handbook on methods for climate change impact assessment and adaptation strategies*. Amsterdam: Institute of Environmental Studies.

Fereres, E., and D.A. Goldhammer. 1990. Deciduous fruit and nut trees in irrigation of agricultural crops. *Agronomy* 30: 987-1017.

Ferrante, J. 2003. *Sociology. A global perspective*. Belmont, CA: Wadsworth/Thompson.

Fischer, G., M. Shah, and H. van Velthuizen. 2001. *Climate change and agricultural vulnerability*. Vienna: International Institute for Applied Systems Analysis.

Flick, U. 1998. *An introduction to qualitative research*. Thousand Oaks, CA: Sage.

Folke, C., S. Carpenter, T. Elmqvist, L. Gunderson, C. Holling, B. Walker, J. Bengtsson, F. Berkes, J. Colding, K. Danell, M. Falkenmark, L. Gordon, R. Kasperson, N. Kautsky, A. Kinzing, S. Levin, K.G. Maler, F. Moberg, L. Ohlsson, P. Olsson, E. Ostrom, W. Reid, J. Rockstrom, H. Savenije, and U. Svedin. 2002. *Resilience and sustainable development: Building adaptive capacity in a world of transformation*. Stockholm: Environmental Advisory Council, Ministry of Environment.

Ford, J., and B. Smit. 2004. A framework for assessing the vulnerability of communities in the Canadian Arctic to risks associated with climate change. *Arctic* 57: 389-400.

Friedmann, H. 1995. Food politics: New dangers, new possibilities. In *Food and agrarian orders in the world economy*, ed. P. McMichael, 15-33. Westport, CT: Praeger.

Funtowicz, S.O., and J.R. Ravetz. 1993. Science for the post-normal age. *Futures* 25: 735-55.

Gertler, M. 1999. Sustainable communities and sustainable agriculture on the Prairies. In *Communities, development, and sustainability across Canada*, ed. J. Pierce and A. Dale, 121-39. Vancouver: UBC Press.

Giorgi, F., and B. Hewison. 2001. Regional climate information – evaluation and projections. In *Climate change 2001: The scientific basis*, ed. J. Houghton, Y. Ding, J. Griggs, M. Noguer, P. van der Linden, and D. Xiaosu, 583-683. Cambridge, UK: Cambridge University Press.

Gitay, H., S. Brown, W. Easterling, and B. Jallow. 2001. Ecosystems and their services. In *Climate change 2001: Impacts, adaptation and vulnerability. Intergovernmental Panel on Climate Change, Working Group II, Third Assessment Report*, ed. J. McCarthy, O. Canziani, N. Leary, D. Dokken, and K. White, 252-70. Cambridge, UK: Cambridge University Press.

Gittel, R.J., and A. Vidal. 1998. *Community organizing: Building social capital as a development strategy.* Thousand Oaks, CA: Sage.

Glantz, M.H. 1991. The use of analogies in forecasting ecological and societal responses to global warming. *Environment* 33: 10-33.

Goodwin, R.B. 1986. Drought: A surface water perspective. In *Drought: The impending crisis?* Proceedings of the Canadian Hydrology Symposium no. 16, Regina, Saskatchewan, 27-43.

Government of Ontario. 2004. Economy. http://stagea.cts.gov.on.ca/MBS/english/about/economy2.html#agriculture (accessed 20 March 2005).

Gregory, J.M., J.F.B. Mitchell, and A.J. Brady. 1997. Summer drought in northern mid-latitudes in a time dependent CO_2 experiment. *Journal of Climate* 10 (4): 662-86.

Gregory, J.N. 1989. *American exodus: The Dust Bowl migration and Okie culture in California.* New York: Oxford University Press.

Gregson, M. 1996. Long term trends in agricultural specialization in the United States: Some preliminary results. *Agricultural History* 70: 90-101.

Gunderson, L.H., and C.S. Holling. 2002. *Panarchy: Understanding transformations in human and natural systems.* Washington, DC: Island Press.

Haas, P., R. Keohane, and M. Levy. 1993. *Institutions for the earth: Sources of effective international environmental protection.* Cambridge, MA: MIT Press.

Hagedorn, R. 1994. *Sociology.* Toronto: Harcourt Brace.

Haimes, Y.Y. 2004. *Risk modeling, assessment, and management.* Hoboken, NJ: Wiley.

Hall, A. 2005. Water: Water and governance. In *Governance for sustainable development: A foundation for the future*, ed. G. Ayre and R. Callway, 118-28. London: Earthscan.

Hamdy, A., M. Abu-Zeid, and C. Lacirignola. 1998. Institutional capacity building for water sector development. *Water International* 23: 126-33.

Hamlet, A.F., and D.P. Lettenmaier. 1999. Effects of climate change on hydrology and water resources in the Columbia River Basin. *Journal of American Water Resource Association* 35: 1597-1624.

Handmer, J.W., S. Dovers, and T.E. Downing. 1999. Societal vulnerability to climate change and variability. *Mitigation and Adaptation Strategies for Global Change* 4: 267-81.

Hanemann, W. 2000. Adaptation and its measurement. *Climatic Change* 45: 571-81.

Harvey, D. 1973. *Social justice and the city.* London: Edward Arnold.

Hengeveld, H.G. 2000. *Climate change digest: Projections for Canada's climate future.* CCD 00-01 special edition. Ottawa: Environment Canada.

Herrington, R., B. Johnston, and F. Hunter. 1997. *Responding to global climate change in the Prairies.* Vol. 3 of *Canada country study: Climate impacts and adaptation.* Downsview, ON: Environment Canada.

Hesch, N.M., and D.H. Burn. 2005. *Analysis of trends in evaporation – phase 1.* A report submitted to Agriculture and Agri-Food Canada, Prairie Farm Rehabilitation Administration.

Hill, H., and J. Vaisey. 1995. Policies for sustainable development. In *Planning for a sustainable future: The case of the North American Great Plains*, ed. D.A. Wilhite and D.A. Wood, 51-62. Proceedings of the symposium, 8-10 May, Lincoln, NE. Lincoln: National Drought Information Centre, University of Nebraska.

Hogg, T., G. Weiterman, and L.C. Tollefson. 1997. *Effluent irrigation: Saskatchewan perspective.* Canada-Saskatchewan Irrigation Diversification Centre (CSIDC), http://www.agr.gc.ca/pfra/csidc/csidpub5_e.htm.

Holling, C.S. 1973. Resilience and stability of ecological systems. *Annual Review of Ecological Systems* 4: 1-23.

–. 1986. The resilience of terrestrial ecosystems: Local surprise and global change. In *Sustainable development of the biosphere*, ed. W.C. Clark and R.E. Munn, 292-317. London: Cambridge University Press.

–. 2001. Understanding the complexity of economic, ecological and social systems. *Ecosystems* 4: 390-405.

Holloway, L.E., and B.W. Ilbery. 1996. Farmer's attitude towards environmental change, particularly global warming, and the adjustment of crop mix and farm management. *Applied Geography* 16: 159-71.

Homer-Dixon, H. 1999. *Environment, scarcity, and violence.* Princeton, NJ: Princeton University Press.

Huq, S., A. Rahman, M. Konate, Y. Sokona, and H. Reid. 2003. *Mainstreaming adaptation to climate change in least developed countries (LDCS).* London: International Institute for Environment and Development.

Huq, S., and H. Reid. 2004. Mainstreaming adaptation in development. *Institute for Development Studies Bulletin* 35 (3): 15-21.

Hutchinson, M.F. 1995. Interpolating mean rainfall using thin plate smoothing splines. *International Journal of Geographic Information Systems* 9: 385-403.

–. 2000. *ANUSPLIN version 4.1 user guide.* Canberra: Centre for Resource and Environmental Studies, Australian National University.

Ilbery, B., and I. Bowler. 1998. From agricultural productivism to postproductivism. In *The geography of rural change,* ed. B. Ilbery, 57-84. London: Longman.

IPCC (Intergovernmental Panel on Climate Change). 2001. *Climate change 2001: The scientific basis. Contribution of Working Group I to the Third Assessment Report of the Intergovernmental Panel on Climate Change,* ed. J. Houghton, Y. Ding, J. Griggs, M. Noguer, P. van der Linden, and D. Xiaosu. Cambridge, UK: Cambridge University Press. Online: http://www.ipcc.ch/.

Ishwaran, K. 1986. *Sociology.* Toronto: Addison-Wesley.

Jamieson, D. 1988. Grappling for a glimpse of the future. In *Societal responses to regional climate change: Forecasting by analogy,* ed. M.H. Glantz, 73-94. Boulder, CO: Westview.

Jones, D. 1987. *Empire of dust.* Calgary: University of Calgary Press.

Jones, R. 2001. An environmental risk assessment/management framework for climate change impact assessments. *Natural Hazards* 23: 197-230.

Jones, R.N. 2000. Analysing the risk of climate change using an irrigation demand model. *Climate Research* 14: 89-100.

Jordan, A., and T. O'Riordan. 1999. Social institutions and climate change: Applying cultural theory and practice. Working paper GEC 97-15. University of East Anglia, UK: Centre for Social and Economic Research on the Global Environment.

Kaiser, H.M., S. Riha, D. Wilkes, D. Rossiter, and R. Sampath. 1993. A farm-level analysis of the economic and agronomic impacts of gradual climate warming. *American Journal of Agricultural Economics* 75: 387-98.

Kandlikar, M., and J. Risbey. 2000. Agricultural impacts of climate change: If adaptation is the answer, what is the question? *Climatic Change* 45: 529-39.

Kane, S., J. Reilly, and J. Tobey. 1992. An empirical study of the economic effects of climate change on world agriculture. *Climatic Change* 21: 17-35.

Kasperson, J.X., and R.E. Kasperson. 2001. *Climate change, vulnerability and social justice.* Stockholm: Stockholm Environment Institute.

Kelly, P.M., and W.N. Adger. 2000. Theory and practice in assessing vulnerability to climate change and facilitating adaptation. *Climatic Change* 47: 325-52.

Ker, A.P., and P. McGowan. 2000. Weather-based adverse selection and the US crop insurance program: The private insurance company perspective. *Journal of Agricultural and Resource Economics* 25 (2): 386-410.

Keskitalo, E.C.H. 2004. A framework for multi-level stakeholder studies in response to global change. *Local Environment* 9 (5): 425-35.

Kling, G., L. Hayhoe, J. Johnson, S. Magnuson, S. Polasky, B. Robinson, M. Shuter, D. Wander, D. Wuebbles, and D. Zak. 2003. *Confronting climate change in the Great Lakes region.* A report of the Union of Concerned Scientists and the Ecological Society of America. http://www.ucsusa.org/greatlakes.

Koshida, G., and W. Avis, eds. 1998. *Canada country study: Climate impacts and adaptation.* Gatineau, QC: Government of Canada. http://www.climatechange.gc.ca/English/publications/ccs/.

Kurukulasuriya, P., and S. Rosenthal. 2003. Climate and change in agriculture: A review of impacts and adaptations. Climate Change Series, Paper No. 91. Washington, DC: World Bank.

Labao, L.M. 1990. *Locality and inequality: Farm and industry structure and socioeconomic conditions*. Albany: The State University of New York Press.

Lac, S. 2004. A climate change adaptation study for the South Saskatchewan River Basin. Institutional Adaptations to Climate Change (IACC) project working paper. http://www.parc.ca/mcri.

Lalonde, K., and B. Corbett. 2004. The state of southern Alberta's water resources. Prepared as background information for the "Confronting Water Scarcity" conference, Lethbridge, Alberta.

Lapp, S., J. Byrne, I. Townshend, and S. Kienzle. 2005. Climate warming impacts on snowpack accumulation in an Alpine watershed. *International Journal of Climatology* 25: 521-36.

Larriviere, J.B., and C.O. Kroncke. 2004. A human capital approach to American Indian earnings: The effects of place of residence and migration. *Social Science Journals* 41: 209-24.

Lawton, G. 2002. Plague of plenty. *New Scientist* 176 (2371): 26.

Leary, N. 1999. A framework for benefit-cost analysis of adaptation to climate change and climate variability. *Mitigation and Adaptation Strategies for Global Change* 4 (3-4): 307-18.

Le Heron, R. 1993 *Globalized agriculture: Political choice*. Oxford: Pergamon Press.

Lee, J.J., D.L. Phillips, and V.W. Benson. 1999. Soil erosion and climate change: Assessing potential impacts and adaptation practices. *Journal of Soil and Water Conservation* 54 (3): 529-36.

Lee, K. 1993. *Compass and the gyroscope*. Washington, DC: Island Press.

Lemmon, D., and F. Warren. 2004. *Climate change impacts and adaptation: A Canadian perspective*. Ottawa: Natural Resources Canada.

Lettenmaier, D.P., A.W. Wood, R.N. Palmer, E.F. Wood, and E.Z. Stakhiv. 1999. Water resources implications of global warming: A US regional perspective. *Climatic Change* 43: 357-579.

Lewandrowski, J., and R. Brazee. 1993. Farm programs and climate change. *Climatic Change* 23: 1-20.

Lewandrowski, J., and D. Schimmelpfennig. 1999. Economic implications of climate change for US agriculture: Assessing recent evidence. *Land Economics* 75 (1): 39-57.

Lim, B., and E. Spanger-Siegfried. 2004. *Adaptation policy frameworks for climate change: Developing strategies, policies and measures*. Cambridge, UK: Cambridge University Press.

Lind, J. 2003. Adaptation, conflict and cooperation in pastoralist East Africa: A case study from South Turkana, Kenya. *Conflict, Security and Development* 3 (3): 315-34.

Loaiciga, H.A., J.B. Valdes, R. Vogel, and J. Garvey. 1996. Global warming and the hydrological cycle. *Journal of Hydrology* 174: 83-127.

Loladze, I. 2002. Rising atmospheric CO_2 and human nutrition: Toward globally imbalanced plant stoichiometry? *Trends in Ecology and Evolution* 17 (10): 457-61.

MacIver, D.C., and F. Dallmeier. 2000. Adaptation to climate change and variability: Adaptive management. *Environmental Monitoring and Assessment* 61: 1-8.

MacKinnon, N., J. Bryden, C. Bell, A. Fuller, and M. Spearman. 1991. Pluriactivity, structural change and farm household vulnerability in Western Europe. *Sociologia Ruralis* 31: 58-71.

Mahul, O., and D. Vermersch. 2000. Hedging crop risk with yield insurance futures and options. *European Review of Agricultural Economics* 27 (2): 109-26.

Major, D.J., W.L. Pelton, C.F. Shaykewich, S.H. Gage, and D.G. Green. 1976. *Heat units for corn in the Prairies*. Canadex factsheet 111.070. Ottawa: Agriculture Canada.

Marsden, T., R. Munton, S. Whatmore, and J. Little. 1986. Towards a political economy of capitalist agriculture: A British perspective. *International Journal of Urban and Regional Research* 10 (4): 498-521.

Martz, D. 2004. *The farmer's share – compare the share 2004*. Centre for Rural Studies and Enrichment, Muenster, Saskatchewan, http://www.nfu.ca/on/misc_files/CTSFINAL%202004.pdf (accessed 23 November 2005).

Massey, D.S., and K.E. Espinosa. 1997. What's driving Mexico-US migration? A theoretical, empirical and policy analysis. *American Journal of Sociology* 102: 939-99.

Mayo, W. 1940. Testimony of Wheeler Mayo, editor of *Sequoyah County Times*, Sallisaw, Oklahoma. In *Select Committee to Investigate the Interstate Migration of Destitute Citizens*, 2122-28. US House of Representatives, 67th Congress, Oklahoma City hearings.

McCarthy, J.J., O.F. Canziani, N.A. Leary, D.J. Dokken, and K.S. White. 2001. *Climate change 2001: Impacts, adaptation and vulnerability. Intergovernmental Panel on Climate Change, Working Group II, Third Assessment Report*. Cambridge, UK: Cambridge University Press.

McCulloch, J.R. 1852. The works of David Ricardo: With a notice of the life and writings of the author. London: J. Murray.

McDean, H.C. 1978. The "Okie" migration as a socio-economic necessity. *Red River Valley Historical Review* 3: 77-92.

McGinn, S.M., A. Toure, O.O. Akinremi, D.J. Major, and A.G. Barr. 1999. Agroclimate and crop response to climate change in Alberta, Canada. *Outlook on Agriculture* 28 (1): 19-28.

McKenney, M.S., W.E. Easterling, and N.J. Rosenberg. 1992. Simulation of crop productivity and responses to climate change in the year 2030: The role of future technologies, adjustments and adaptations. *Agricultural and Forest Meteorology* 59 (1-2): 103-27.

McKenney, D.W., M.F. Hutchinson, J.L. Kesteven, and L.A. Venier. 2001. Canada's plant hardiness zones revisited using modern climate interpolation techniques. *Canadian Journal of Plant Science* 81: 129-43.

McLeman, R., and B. Smit. 2006. Migration as adaptation to climate change. *Climatic Change* 76 (1-2): 31-53.

McMillan, R.T. 1936. Some observations on Oklahoma population movements since 1930. *Rural Sociology* 1: 332-43.

McNeil, R. 2004. Costs of adaptation options. In *Expanding the dialogue on climate change and water management in the Okanagan Basin, British Columbia*, ed. S. Cohen, D. Neilsen, and R. Welbourn, 161-63. Project A463/433, submitted to Adaptation Liaison Office, Climate Change Action Fund, Natural Resources Canada, Ottawa. http://www.ires.ubc.ca.

McRae, T., C.A.S. Smith, and L.J. Gregorich, eds. 2000. *Environmental sustainability of Canadian agriculture: Report of the Agri-Environmental Indicator Project. A summary*. Ottawa: Agriculture and Agri-Food Canada.

McWilliams, C. 1942. *Ill fares the land: Migrants and migratory labor in the United States*. Boston: Little, Brown.

Mearns, L.O., F. Giorgi, L. McDaniel, and C. Shields. 2003. Climate scenarios for the south eastern US based on GCM and regional model simulations. *Climatic Change* 60: 7-35.

Mearns, L.O., C. Rosenzweig, and R. Goldberg. 1992. Effects of changes in interannual climatic variability on CERES-wheat yields: Sensitivity and 2 ´ CO_2 General Circulation Model scenarios. *Agricultural and Forest Meteorology* 62: 159-89.

Mehlman, S. 2003. *Historical and projected temperature and precipitation trends in the Annapolis Valley, Nova Scotia*. Technical report for the Annapolis Valley Climate Change Policy Response Pilot Project, Clean Annapolis River Project.

Meinert, B. 2003. Presentation for "Meeting the Challenges of Climate Change" roundtable session sponsored by C-CIARN Agriculture and C-CIARN Prairies at Grainworld 2003, Winnipeg.

Mendelsohn, R., W. Nordhaus, and D. Shaw. 1994. The impact of global warming on agriculture: A Ricardian analysis. *American Economic Review* 84: 753-71.

Merritt, W., and Y. Alila. 2004. Hydrology. In *Expanding the dialogue on climate change and water management in the Okanagan Basin, British Columbia*, ed. S. Cohen, D. Neilsen, and R. Welbourn, 63-86. Project A463/433, submitted to Adaptation Liaison Office, Climate Change Action Fund, Natural Resources Canada, Ottawa. http://www.ires.ubc.ca.

Merritt, W., Y. Alila, M. Barton, B. Taylor, S. Cohen, and D. Neilsen. 2006. Hydrologic response to scenarios of climate change in the Okanagan Basin, British Columbia. *Journal of Hydrology* 326: 79-108.

Miles, E.L., A.F. Hamlet, A.K. Snover, B. Callahan, and D. Fluharty. 2000. Pacific Northwest regional assessment: The impacts of climate variability and climate change on the water resources of the Columbia River Basin. *Journal of the American Water Resources Association* 36: 399-420.

Morduch, J., and M. Sharma. 2002. Strengthening public safety nets from the bottom-up. *Development Policy Review* 20 (5): 569-88.

Morrison, J., M.C. Quick, and M.G.G. Foreman. 2002. Climate change in the Fraser River watershed: Flow and temperature projections. *Journal of Hydrology* 263: 230-44.

Moss, S., C. Pahl-Wostl, and T. Downing. 2001. Agent-based integrated assessment modelling: The example of climate change. *Integrated Assessment* 2 (1): 17-30.

Mote, P., et al. 1999. *Impacts of climate variability and change – Pacific Northwest. A report of the Pacific Northwest regional assessment group for the US Global Change Research Program.* JISAO/SMA Climate Impacts Group, JISAO contribution no. 715. Seattle: University of Washington.

Mote, P.W., E.A. Parson, A.F. Hamlet, W.S. Keeton, D. Lettenmaier, N. Mantua, E.L. Miles, D.W. Peterson, D.L. Peterson, R. Slaughter, and A.K. Snover. 2003. Preparing for climate change: The water, salmon, and forests of the Pacific Northwest. *Climatic Change* 61: 45-88.

Mueser, P.R., and P.E Graves. 1995. Examining the role of economic opportunity and amenities in explaining population redistribution. *Journal of Urban Economics* 37: 176-200.

Myers, S.M. 2000. The impact of religious involvement in migration. *Social Forces* 79: 755-83.

Nagy, C. 2001. *Agriculture energy use of adaptation options to climate change.* Report to the Prairie Adaptation Research Collaborative (PARC).

Nakicenovic, N., J. Alcamo, G. Davis, B. deVries, J. Fenhann, S. Gaffin, K. Gregory, A. Grubler, T.Y. Jung, T. Kram, E.L. La Rovere, L. Michaelis, S. Mori, T. Morita, W. Papper, H. Pitcher, L. Price, K. Riahi, A. Roehrl, H.-H. Rogner, A. Sankovski, M. Schlesinger, P. Shukla, S. Smith, R. Swart, S. van Rooijen, N. Victor, and Z. Dadi. 2000. *Emissions scenarios. A special report of Working Group III of the Intergovernmental Panel on Climate Change.* Cambridge, UK: Cambridge University Press.

National Farmers Union (NFU). 2005. *Solving the farm crisis: A sixteen-point plan for Canadian farm and food security.* Saskatoon: National Farmers Union. Online: http://www.nfu.ca/ briefs/Ten_point_plan_to_end_farm_crisis_EIGHTEEN_-_FINAL_bri.pdf

–. 2003. Climate change in Canada: Adaptation and mitigation. A brief to the Standing Senate Committee on Agriculture and Forestry. 13 February. http://www.nfu.ca.

Natural Resources Canada. 2002. *Climate change impacts and adaptation: A Canadian perspective.* Ottawa: Agriculture Climate Change Impacts and Adaptation Directorate.

–. 2004. *Atlas of Canada.* http://atlas.gc.ca/site/english/.

Nee, V., and J. Sanders. 2001. Understanding the diversity of migrant incorporation: A forms-of-capital approach. *Ethnic and Racial Studies* 24: 386-411.

Neilsen, D., W. Koch, W. Merritt, G. Frank, S. Smith, Y. Alila, J. Carmichael, T. Neale, and R. Welbourn. 2004b. Risk assessment and vulnerability – case studies of water supply and demand. In *Expanding the dialogue on climate change and water management in the Okanagan Basin, British Columbia,* ed. S. Cohen, D. Neilsen, and R. Welbourn, 115-35. Project A463/ 433, submitted to Adaptation Liaison Office, Climate Change Action Fund, Natural Resources Canada, Ottawa. http://www.ires.ubc.ca.

Neilsen, D., W. Koch, S. Smith, and G. Frank. 2004a. Crop water demand scenarios for the Okanagan Basin. In *Expanding the dialogue on climate change and water management in the Okanagan Basin, British Columbia,* ed. S. Cohen, D. Neilsen, and R. Welbourn, 89-112. Project A463/433, submitted to Adaptation Liaison Office, Climate Change Action Fund, Natural Resources Canada, Ottawa. http://www.ires.ubc.ca.

Neilsen, D., S. Smith, G. Frank, W. Koch, Y. Alila, W. Merritt, B. Taylor, M. Barton, J. Hall, and S. Cohen, 2006. Potential impacts of climate change on water availability for crops in the Okanagan Basin, British Columbia. *Canadian Journal of Soil Science* 86: 921-35.

Neilsen D., C.A.S. Smith, W. Koch, G. Frank, J. Hall, and P. Parchomchuk. 2001. *Impact of climate change on irrigated agriculture in the Okanagan Valley, British Columbia.* Final report, Climate Change Action Fund Project A087. Ottawa: Natural Resources Canada.

Neilsen, D., S. Smith, W. Koch, J. Hall, and P. Parchomchuk. 2002. *Impact of climate change on crop water demand and crop suitability in the Okanagan Valley, British Columbia*. Final report submitted to the Climate Change Action Fund, Ottawa.

Neilsen, G.H., E.J. Hogue, T. Forge, and D. Neilsen. 2003. Surface application of mulches and biosolids affect orchard soil properties after 7 years. *Canadian Journal of Soil Science* 83: 131-37.

Nemanishen, W. 1998. *Drought in the Palliser Triangle: A provisional primer*. Regina: Prairie Farm Rehabilitation Administration.

Network of Concerned Farmers. 2005. Why do genetically modified crops perform worse in drought? Press release, http://www.non-gm-farmers.com/news_details.asp?ID=2253.

Newman, D. 2004. *Sociology: Exploring the architecture of everyday life*. Thousand Oaks, CA: Pine Forge Press.

Nordhaus, W., and J. Boyer. 2000. *Warming the world: Economic models of global warming*. Cambridge, MA: MIT Press.

Northwest Hydraulic Consultants. 2001. *Hydrology, water use and conservation flows for kokanee salmon, and rainbow trout in the Okanagan Lake Basin, BC*. Report to British Columbia Fisheries. Victoria.

Nyirfa, W.N., and B. Harron. 2001. *Assessment of climate change on the agricultural resources of the Canadian Prairies*. Report to the Prairie Adaptation Research Collaborative on PARC project QS-3.

O'Brien, K.L., and R.M. Leichenko. 2000. Double exposure: Assessing the impacts of climate change within the context of economic globalization. *Global Environmental Change* 10: 221-32.

Office of the Auditor General. 2004. *Eighth annual report of the Commissioner of the Environment and Sustainable Development*. http://www.oagbvg.gc.ca/domino/reports.nsf/html/c2004menu_e.html.

Okanagan Valley Tree Fruit Authority. 1995. *Tree fruit suitability in the Okanagan, Similkameen, and Creston valleys. Technical reference manual*. Summerland: Okanagan Valley Tree Fruit Authority.

Olmstead, C.W. 1970. The phenomena, functioning units and systems of agriculture. *Geographica Polonica* 19: 31-41.

Olsson, P., C. Folke, and T. Hahn. 2004. Social-ecological transformation for ecosystem management: The development of adaptive co-management of a wetland landscape in southern Sweden. *Ecology and Society* 9 (4): 2.

Owensby, C., R. Cochran, and L. Auen. 1996. Effects of elevated carbon dioxide on forage quality for ruminants. In *Carbon dioxide, populations, and communities*, ed. C. Koerner and F. Bazzaz, 363-71. New York: Elsevier.

Pahl-Wostl, C. 2002. Participative and stakeholder-based policy design, evaluation and modeling processes. *Integrated Assessment* 3 (1): 3-14.

Palys, T. 1997. *Research decisions: Quantitative and qualitative perspectives*. Toronto: Harcourt Brace Jovanovich.

Parry, M.L. 1990. *Climate change and world agriculture*. London: Earthscan.

Parry, M., N. Arnell, M. Hulme, R. Nicholls, and M. Livermore. 1998. Adapting to the inevitable. *Nature* 395: 741-42.

Pattey, E., I.B. Strachan, J.B. Boisvert, R.L. Desjardins, and N.B. McLaughlin. 2001. Detecting effects of nitrogen rate and weather on corn growth using micrometeorological and hyperspectral reflectance measurements. *Agriculture and Forest Meteorology* 108 (2): 85-164.

Patton, A. 2003. Presentation to Standing Senate Committee on Agriculture and Forestry hearings, held in Vancouver, 28 February, 2003 (transcript).

Payne, J.T., A.W. Wood, A.F. Hamlet, R.N. Palmer, and D.P. Lettenmaier. 2004. Mitigating the effects of climate change on the water resources of the Columbia River Basin. *Climatic Change* 62: 233-56.

Pelletier, D.L., V. Kraak, C. McCullum, U. Unsitalo, and R. Rich. 1999. Community food security: Salience and participation at the community level. *Agriculture and Human Values* 16 (4): 401-19.

Pettapiece, W.W., ed. 1995. *Land suitability rating system for agricultural crops: 1. Spring-seeded small grains.* Technical bulletin 1995-6E. Ottawa: Research Branch, Agriculture and Agri-Food Canada.

PFRA (Prairie Farm Rehabilitation Authority). 2000. *Prairie agricultural landscapes: A land resource review.* Regina: PFRA.

–. 2003. *Watch for sulphates and blue-green algae in cattle water supplies.* http://www.agr.gc.ca/pfra/drought/article_e.htm.

Pimental, D., P. Hepperly, J. Hanson, D. Douds, and R. Seidel. 2005. Environmental, energetic, and economic comparisons of organic and conventional farming systems. *BioScience* 55 (7): 573-82.

Pittock, B., and R. Jones. 2000. Adaptation to what and why? *Environmental Monitoring Assessment* 61: 9-35.

Postel, S. 2000. Entering an era of water scarcity: The challenges ahead. *Ecological applications* 10: 941-48.

Putnam, R.D. 1995. Bowling alone: America's declining social capital. *Journal of Democracy* 6: 65-78.

Quick, M.C. 1995. The HBV model. In *Computer models of watershed hydrology,* ed. V.P. Singh, 223-80. Highlands Ranch, CO: Water Resources Publications.

Rausser, G., and E. Hochman. 1979. *Dynamic agricultural systems.* New York: North Holland.

Rees, H., E. Chow, P. Loro, J. Lavoie, J. Monteith, and A. Blaauw. 2002. Hay mulching to reduce runoff and soil loss under intensive potato production in northwestern New Brunswick, Canada. *Canadian Journal of Soil Science* 82: 249-58.

Reid, S. 2003. *Winds of change: Farm-level perception and adaptation to climate risk in Perth County, Ontario.* M.Sc. thesis, School of Rural Planning and Development, University of Guelph.

Reid, S., B. Smit, W. Caldwell, and S. Belliveau. 2007. Vulnerability and adaptation to climate risks in Ontario agriculture. *Mitigation and Adaptation Strategies for Global Change* 12 (4): 609-37.

Reilly, J. 1995. Climate change and global agriculture: Recent findings and issues. *American Journal of Agricultural Economics* 77: 727-33.

–. 2001. *Agriculture: The potential consequences of climate variability and change for the United States.* New York: Cambridge University Press.

Reilly, J., F. Tubiello, B. McCarl, D. Abler, R. Darwin, K. Fuglie, S. Hollinger, C. Izaurralde, S. Jagtap, J. Jones, L. Mearns, D. Ojima, E. Paul, K. Paustian, S. Riha, N. Rosenberg, and C. Rosenzweig. 2003. US agriculture and climate change: New results. *Climatic Change* 57: 43-69.

Reinsborough, M. 2003. A Ricardian model of climate change in Canada. *Canadian Journal of Economics* 36: 21-40.

Ripley, E.A. 1988. *Drought prediction of the Canadian Prairies.* Canadian Climate Centre report no. 88-4. Saskatoon: National Hydrology Research Centre.

Risbey, J., M. Kandlikar, H. Dowlatabadi, and D. Graetz. 1999. Scale, context, and decision making in agricultural adaptation to climate variability and change. *Mitigation and Adaptation Strategies for Global Change* 4: 137-65.

Rochester, A. 1940. *Why farmers are poor.* 1975 ed. New York: Arno Press.

Roderick, M.L., and G.D. Farquhar. 2002. The cause of decreased pan evaporation over the past 50 years. *Science* 298: 1410-11.

Rosenberg, N. 1981. Technologies and strategies in weatherproofing crop production. In *Climate impact on food supplies: Strategies and technologies for climate-defensive food production,* ed. L.E. Slater and S.K. Levins, 157-80. Boulder, CO: Westview Press.

Rosenberg, N.J., M.S. McKenney, W.E. Easterling, and K.M. Lemon. 1992. Validation of EPIC model simulation of crop responses to current climate and CO_2 conditions: Comparisons with census, expert judgement and experimental plot data. *Agricultural and Forest Meteorology* 59: 35-51.

Rosenzweig, C. 1985. Potential CO_2-induced climate effects on North American wheat-producing regions. *Climatic Change* 7: 367-89.

Rosenzweig, C., A. Iglesias, X. Yang, P. Epstein, and E. Chivian. 2000. *Climate change and US agriculture: The impacts of warming and extreme weather events on productivity, plant diseases and pests.* Boston: Center for Health and the Global Environment, Harvard Medical School.

Rosenzweig, C., and M.L. Parry. 1994. Potential impacts of climate change on world food supply. *Nature* 367: 133-38.

Ruddiman, W. 2005. How did humans first alter global climate? *Scientific American* 292: 46-54.

Ryan, K., and L. Destefano. 2000. *Evaluation in a democratic society: Deliberation, dialogue and inclusion.* San Francisco: Jossey-Bass.

Sanderson, D. 2000. Cities, disasters and livelihoods. *Risk Management: An International Journal* 2 (4): 49-58.

Sauchyn, D. 2003. Information in transcript for Standing Senate Committee on Agriculture and Forestry hearings. 4 February 2003 (transcript).

Sauchyn, D.J., and A.B. Beaudoin. 1998. Recent environmental change in the southwestern Canadian Plains. *Canadian Geographer* 42 (4): 337-53.

Sauchyn, D.J., E. Barrow, R.F. Hopkinson, and P. Leavitt. 2002. Aridity on the Canadian Plains. *Géographie physique et Quaternaire* 56 (2-3): 247-59.

Sauchyn, D.J., S. Kennedy, and J. Stroich. 2005. Drought, climate change and the risk of desertification on the Canadian Plains. *Prairie Forum* 30 (1): 143-56.

Sauchyn, D.J., J. Stroich, and A. Beriault. 2003. A paleoclimatic context for the drought of 1999-2001 in the northern Great Plains. *Geographical Journal* 169 (2): 158-67.

Schindler, D.W. 2001. The cumulative effects of climate warming and other human stresses on Canadian freshwaters in the new millennium. *Canadian Journal of Fish Aquatic Science* 58: 18-29.

Schneider, S.H., W.E. Easterling, and L.O. Mearns. 2000. Adaptation: Sensitivity to natural variability, agent assumption and dynamic climate changes. *Climatic Change* 45: 203-31.

Schröter, D., C. Polsky, and A.G. Patt. 2005. Assessing vulnerabilities to the effects of global change: An eight step approach. *Mitigation and Adaptation Strategies for Global Change* 10 (4): 573-95.

Schweger, C., and C. Hooey. 1991. Climate change and the future of Prairie agriculture. In *Alternative futures for Prairie agricultural communities,* ed. J. Martin, 1-36. Edmonton: University of Alberta.

Scoones, I., 2004. Climate change and the challenge of non-equilibrium thinking. *IDS Bulletin* 35 (3): 114-19.

Semenov, N.A., and E.M. Barrow. 1997. Use of a stochastic weather generator in the development of climate change scenarios. *Climatic Change* 35: 397-414.

Semenov, N.A., and J.R. Porter. 1995. Climatic variability and the modelling of crop yields. *Agricultural and Forest Meteorology* 73: 265-83.

Shepherd, P., J. Tansey, and H. Dowlatabadi. 2006. Context matters: What shapes adaptation to water stress in the Okanagan? *Climatic Change* 78: 31-62.

Shields, J.A., and W.K. Sly. 1984. *Aridity indices derived from soil and climatic parameters.* Technical bulletin 1984-14E. Ottawa: Research Branch, Agriculture Canada.

Shortt, R., W. Caldwell, J. Ball, and P. Agnew. 2004. *A participatory approach to water management: Irrigation advisory committees in southern Ontario.* Paper presented at 57th Canadian Water Resources Association annual congress, Montréal, Québec.

Singh, B., C.R. Bryant, P. André, J.-P. Thouez, M. El Mayaar, L. Huberdeau, and D. Provençal. 1995. Impacts potentiels du changement climatique dû à une hausse atmosphérique de gaz à effet de serre sur l'agriculture au Québec. In *1er Colloque de l'Université de Montréal sur l'environnement: Analyse et intervention,* ed. C.E. Delisle, M.A. Bouchard, P. André, and J. Zayed, 465-76. Collection Environnement de l'Université de Montréal 18.

Singh, B., M. El Maayar, P. André, C.R. Bryant, and J.-P. Thouez. 1998. Impacts of a GHG-induced climate change on crop yields: Effects of acceleration in maturation, moisture stress and optimal temperature. *Climatic Change* 38: 51-86.

Singh, B., M. El Mayaar, P. André, J.-P. Thouez, C.R. Bryant, and D. Provençal. 1996. Influence d'un changement climatique dû à une hausse de gaz à effet de serre sur l'agriculture au Québec. *Atmosphère-Océan* 34 (2): 379-99.

Singh, B., and R.B. Stewart. 1991. Potential impacts of a CO_2-induced climate change using the GISS Scenario on agriculture in Québec, Canada. *Agriculture, Ecosystems and Environment* 35: 327-47.

Smit, B. 1991. Decisions in agriculture in the face of uncertainty: Role of climate impact assessment. In *Changing climate in relation to sustainable agriculture*, ed. P. Dzikowski. Fredericton, NB: Canadian Society of Agrometeorology.

–. 1993. *Adaptation to climatic variability and change.* Guelph: Environment Canada.

Smit, B., R. Blain, and P. Keddie. 1997. Corn hybrid selection and climatic variability: Gambling with nature? *Canadian Geographer* 41: 429-38.

Smit, B., M. Brklacich, R.B. Stewart, R. McBride, M. Brown, and D. Bond. 1989. Sensitivity of crop yields and land resource potential to climatic change in Ontario, Canada. *Climatic Change* 14 (2): 153-74.

Smit, B., I. Burton, R.J.T. Klein, and J. Wandel. 2000b. An anatomy of adaptation to climate change and variability. *Climatic Change* 45: 223-51.

Smit, B., E. Harvey, and C. Smithers. 2000a. How is climate change relevant to farmers? In *Climate change communication: Proceedings of an international conference*, ed. D. Scott, B. Jones, J. Audrey, R. Gibson, P. Key, L. Mortsch, K. Warriner. Hull, QC: Environment Canada.

Smit, B., L. Ludlow, and M. Brklacich. 1988. Implications of a global climatic warming for agriculture: A review and appraisal. *Journal of Environmental Quality* 17 (4): 519-27.

Smit, B., D. McNabb, and J. Smithers. 1996. Agricultural adaptation to climatic variation. *Climatic Change* 33: 7-29.

Smit, B., and O. Pilifosova. 2003. From adaptation to adaptive capacity and vulnerability reduction. In *Climate change, adaptive capacity and development*, ed. J. Smith, R. Klein, and S. Huq, 9-28. London: Imperial College Press.

Smit, B., and W. Skinner. 2002. Adaptation options in agriculture to climate change: A typology. *Mitigation and Adaptation Strategies for Global Change* 7: 85-114.

Smit, B., and J. Wandel. 2006. Adaptation, adaptive capacity, and vulnerability. *Global Environmental Change* 16: 282-92.

Smith, A. 1838. *An inquiry into the nature and causes of the wealth of nations.* Edinburgh: Thomas Nelson.

Smith, J., B. Lavender, H. Auld, D. Broadhurst, and T. Bullock. 1998. *Adapting to climate change and variability in Ontario.* Vol. 4 of *Canada country study: Climate impacts and adaptation.* Downsview, ON: Environment Canada.

Smith, J.B., and S.S. Lenhart. 1996. Climate change adaptation policy options. *Climate Research* 6: 193-201.

Smith, K., C.B. Barrett, and P.W. Box. 2000. Participatory risk mapping for targeting research and assistance, with an example from East African pastoralists. *World Development* 28 (11): 1945-59.

Smithers, J., and A. Blay-Palmer. 2001. Technology innovation as a strategy for climate adaptation in agriculture. *Applied Geography* 21: 175-97.

Smithers, J., and B. Smit. 1997a. Human adaptation to climatic variability and change. *Global Environmental Change* 7 (2): 129-46.

–. 1997b. Agricultural system response to environmental stress. In *Agricultural restructuring and sustainability*, ed. B. Ilbery, Q. Chiotti, and T. Rickard, 167-84. London: CAB International.

Sobool, D., and S. Kulshreshtha. 2003. *Socio-economic database: South Saskatchewan River Basin (Saskatchewan and Alberta).* Report prepared for the Prairie Adaptation Research Collaborative, Department of Agricultural Economics, University of Saskatchewan.

Social Dimensions of Climate Change Working Group. 2005. *Social dimensions of the impact of climate change on water supply and use in the city of Regina.* Regina: Canadian Plains Research Center.

Southern, J.H. 1939. *Farm tenancy in Oklahoma.* Report no. B-239. Stillwater: Oklahoma Agricultural Experiment Station.

Southworth, J., J.C. Randolph, M. Habeck, O.C. Doering, R.A. Pfeifer, D.G. Rao, and J.J. Johnston. 2000. Consequences of future climate change and changing climate variability on maize yields in the midwestern United States. *Agriculture, Ecosystems and Environment* 82: 139-58.

SSCAF (Senate Standing Committee on Agriculture and Forestry). 2003. *Climate change: We are at risk.* Ottawa: SSCAF.

Stanley, J. 1992. *Children of the Dust Bowl: The true story of the school at Weedpatch Camp.* New York: Crown Publishers.

Statistics Canada. 2000. *Farm financial survey,* Cat. no. 21-Foo8-XIB.

–. 2003. *The Daily,* 26 August. Online: http://www.statcan.ca/Daily/English/030826/d030826a.htm.

–. 2004. Total income of farm operators. *The Daily,* 20 July.

–. 2005a. *CANSIM data base: Canadian socio-economic information management system* [computer file]. Ottawa.

–. 2005b. *Census of agriculture 2001.* http://www.statcan.ca (accessed 1 June 2005).

Stein, W.J. 1973. *California and the Dust Bowl migration.* Westport, CT: Greenwood Press.

Stewart, A., and T. O'Brien. 2001. *Application of long range seasonal climate forecasts to improve adaptation to climate variability and change in agriculture and water sectors of the Canadian Prairies.* Report to the Prairie Adaptation Research Collaborative on PARC project QS-5.

Stewart, I.T., D.R. Cayan, and M.D. Dettinger. 2004. Changes in snowmelt runoff timing in western North America under a "business as usual" climate change scenario. *Climatic Change* 62: 217-32.

Stewart, R.B. 1983. *Modelling methodology for assessing crop production potentials in Canada.* Contribution 1983-12E. Ottawa: Research Branch, Agriculture Canada.

Stratos. 2003. *Building confidence: Corporate sustainability reporting in Canada.* Ottawa: Stratos Inc. and Alan Willis and Associates.

Stroh Consulting. 2005. *Agriculture adaptation to climate change in Alberta: Focus group results.* Report prepared for Alberta Agriculture, Food and Rural Development, Edmonton.

Strzepek, K., D. Major, C. Rosenzweig, A. Iglesias, D. Yates, A. Holt, and D. Hillel. 1999. New methods of modeling water availability for agriculture under climate change: The US cornbelt. *Journal of the American Water Resources Association* 35 (6): 1639-55.

Sutherland, K., B. Smit, V. Wulf, and T. Nakalevu. 2005. Vulnerability in Samoa. *Tiempo* 54: 11-15.

Taeuber, C., and C.S. Hoffman. 1937. *Recent migration from the drought areas.* Washington, DC: US Resettlement Administration, US Farm Security Administration, US Bureau of Agricultural Economics.

Tansey, J., and S. Langsdale. 2004. Exploring anticipatory adaptation in the Okanagan, BC. In *Expanding the dialogue on climate change and water management in the Okanagan Basin, British Columbia,* ed. S. Cohen, D. Neilsen, and R. Welbourn, 165-74. Project A463/433, submitted to Adaptation Liaison Office, Climate Change Action Fund, Natural Resources Canada, Ottawa. http://www.ires.ubc.ca.

Taylor, B., and M. Barton. 2004. Climate change scenarios. in Cohen, S., D. Neilsen, and R. Welbourn (eds.). *Expanding the dialogue on climate change & water management in the Okanagan Basin, British Columbia.* Project A463/433, submitted to Adaptation Liaison Office, Climate Change Action Fund, Natural Resources Canada, Ottawa. In *Expanding the dialogue on climate change and water management in the Okanagan Basin, British Columbia,* ed. S. Cohen, D. Neilsen, and R. Welbourn, 47-53. Project A463/433, submitted to Adaptation Liaison Office, Climate Change Action Fund, Natural Resources Canada, Ottawa. http://www.ires.ubc.ca.

Thomson, A.M., R.A. Brown, N.J. Rosenberg, R.C. Izaurralde, and V. Benson. 2005. Climate change impacts for the conterminous USA: An integrated assessment. Part 3: Dryland production of grain and forage crops. *Climatic Change* 69: 43-65.

Thorleifson, H., H. Maathuis, and J. Lebedin. 2001. *Potential impact of climate change on Prairie groundwater supplies: Review of current knowledge.* Report to the Prairie Adaptation Research Collaborative on PARC Project PARC-09.

Timmerman, P. 1989 *Everything else will not remain equal: The challenge of social research in the face of global climate warming.* Report of the First US-Canada Symposium on Impacts of Climate Change on the Great Lakes Basin. Joint report no. 1, National Climate Program Office/Canadian Climate Centre.

Tol, R., S. Fankhauser, and J. Smith. 1998. The scope for adaptation to climate change: What can we learn from the impact literature? *Global Environmental Change* 8: 109-23.

Tollenaar, M., L.M. Dwyer, and D. E. McCullough. 1994. Physiological basis of the genetic improvement of corn. In *Genetic improvement of field crops*, ed. G.A. Slater, 183-236. New York: Marcel Dekker.

Tollenaar, M., and J. Wu. 1999. Yield improvement in temperate maize is attributable to greater stress tolerance. *Crop Science* 39: 1597-1604.

Toth, F.L. 1999. *Fair weather? Equity concerns in climate change.* London: Earthscan.

Troughton, M.J. 1991. Agriculture and rural resources. In *Resource management and development*, ed. B. Mitchell, 53-84. Toronto: Oxford University Press.

Turner, B.L., R.E. Kasperson, P.A. Matson, J.J. McCarthy, R.W. Corell, L. Christensen, N. Eckley, J.X. Kasperson, A. Luers, M.L. Martello, C. Polsky, A. Pulsipher, and A. Schiller. 2003. A framework for vulnerability analysis in sustainability science. *Proceedings of the National Academy of Sciences* 100 (14): 8074-79.

Turton, A.R. 1999. Water scarcity and social adaptive capacity: Towards an understanding of the social dynamics of water demand management in developing countries. MEWREW occasional paper no. 9. London: University of London School of African Studies.

Turvey, C. 2001. Weather derivatives for specific event risks in agriculture. *Review of Agricultural Economics* 23 (2): 333-51.

Tyrchniewicz, A., and Q. Chiotti. 1997. Agriculture and climate change: A Prairie perspective, Appendix A. In *Responding to global climate change in the Prairies*. Vol. 3 of *Canada country study: Climate impacts and adaptation*. Downsview, ON: Environment Canada.

Tyrchniewicz, E. 2003. Submission to Standing Senate Committee on Agriculture and Forestry, Ottawa, 20 February.

UNEP (United Nations Environment Program). 1994. *United Nations convention to combat desertification in those countries experiencing drought and/or desertification, particularly in Africa.* Geneva.

UNFCC (United Nations Framework Convention on Climate Change). 1992. *United Nations framework convention on climate change: Text.* Geneva: World Meteorological Organization and United Nations Environment Program.

USDA (United States Department of Agriculture). 1999. *Managing risk in farming: Concepts, research and analysis.* Agricultural economics report no. 774, Economic Research Service. http://www.ers.usda.gov/publications/aer774/.

US National Research Council. 1989. *Alternative agriculture.* Washington, DC: National Academy Press.

Vaisey, J.S., T.W. Weins, and R.J. Wettlaufer. 1996. The permanent cover program – is twice enough? Paper presented at Soil and water conservation policies: Successes and failures, Prague, Czech Republic, 17-20 September. Document available through Agriculture and Agri-Food Canada. http://www.arg.gc.ca/pfra/pub/pcpaper2.htm.

van Herk, J. 2001. Presentation in summary report for "Risks and Opportunities from Climate Change for the Agricultural Sector" workshop sponsored by Farming Systems Research, University of Guelph, March. Document available through C-CIARN Agriculture. http://www.c-ciarn.uoguelph.ca/documents/2001_workshop_report.pd.

Walker, B., S.R. Carpenter, J. Anderies, N. Abel, G.S. Cumming, M. Janssen, L. Lebel, J. Norberg, G.D. Peterson, and R. Pritchard. 2002. Resilience management in social-ecological systems: A working hypothesis for a participatory approach. *Conservation Ecology* 6 (1). Online: http://www.ecologyandsociety.org/vol6/iss1/art14/print.pdf.

Wall, E., and B. Smit. 2005. Climate change adaptation in light of sustainable agriculture. *Journal of Sustainable Agriculture* 27 (1): 113-23.

Wall, E., B. Smit, and J. Wandel. 2004. *Canadian agri-food sector adaptation to risks and opportunities from climate change: Position paper on climate change, impacts, and adaptation in Canadian agriculture.* Guelph: C-CIARN Agriculture.

Wall, E., and K. Marzall. 2006. Adaptive capacity for climate change in Canadian rural communities. *Local Environment* 11 (4): 373-97.

Wandel, J., and B. Smit. 2000. Agricultural risk management in light of climate variability and change. In *Agriculture and environmental sustainability in the new countryside*, ed. H. Millward, K. Beesley, B. Ilbery, and L. Harrington, 30-39. Winnipeg: Hignell Printing.

Ward, N. 1993. The agricultural treadmill and the rural environment in the post-productivist era. *Sociologia Ruralis* 33: 348-64.

Watts, M.J., and H.G. Bohle. 1993. The space of vulnerability: The causal structure of hunger and famine. *Progress in Human Geography* 17: 43-67.

Weaver, A. 2004. The science of climate change. In *Hard choices: Climate change in Canada*, ed. H. Coward and A. Weaver, 13-43. Waterloo, ON: Wilfrid Laurier University Press.

Weber, M., and G. Hauser. 2003. A regional analysis of climate change impacts on Canadian agriculture. *Canadian Public Policy* (June): 163-79.

Wellstead, A., D. Davidson, and R. Stedman. 2002. *Assessing the potential for policy responses to climate change*. Final report to the Prairie Adaptation Research Collaborative on PARC Project 011. http://www.parc.ca.

Wenger, R., B. Hallett, J. Harris, and D.S. DeVault. 2000. An assessment of ecosystem risks in the St. Croix National Scenic Riverway. *Environmental Management* 25 (6): 599-611.

Wheaton, E., V. Wittrock, S. Kulshreshtha, G. Koshida, C. Grant, A. Chipanshi, and B. Bonsal. 2005. *Lessons learned from the Canadian drought years of 2001 and 2002: Synthesis report for Agriculture and Agri-Food Canada*. SRC publication no. 11602-46E03. Saskatoon: Saskatchewan Research Council.

Wheaton, E.E., and D.C. McIver. 1999. A framework and key questions for adapting to climate variability and change. *Mitigation and Adaptation Strategies for Global Change* 4: 215-25.

Wilbanks, T., and R.W. Kates. 1999. Global change in local places: How scale matters. *Climatic Change* 43: 601-28.

Wilby, R.L., C.W. Dawson, and E.M. Barrow. 2002. SDSD+M – a decision support tool for the assessment of regional climate change impacts. *Environmental and Modelling Software* 17: 145-57.

Wilby, R.L., and T.M.L. Wigley. 1997. Downscaling general circulation model output: A review of methods and limitations. *Progress in Physical Geography* 21: 530-48.

Wilhite, D.A. 2005. Drought management: Shifting the paradigm from crisis to risk management. In *The science, impacts and monitoring of drought in western Canada: Proceedings of the 2004 Prairie Drought Workshop*, ed. D. Sauchyn et al, 1-9. Regina: CPRC Press.

Willems, S. 2004. *Institutional capacity and climate actions: Summary paper*. COM/ENV/EPOC/IEA/SLT(2004)2. Paris: Environment Directorate, Organization for Economic Cooperation and Development.

Williams, G.D.V., R.A. Fautley, K.H. Jones, R.B. Stewart, and E.E. Wheaton. 1988. *Estimating effects of climate change on agriculture in Saskatchewan, Canada. International Institute for Applied Systems Analysis*. Reprinted from *The impact of climatic variations on agriculture. Volume 1: Assessments in cool temperate and cold regions*, ed. M.L. Parry, T.R. Carter, and N.T. Konijn. Dordrecht, Netherlands: Reidel.

Williams, G.D.V., and E.E. Wheaton. 1998. Estimating biomass and wind erosion impacts for several climatic scenarios: A Saskatchewan case study. *Prairie Forum* 23 (1): 49-66.

Williams, P., M. Paridaen, K. Dossa, and M. Dumais. 2001. *Agritourism market and product development status report*. Burnaby, BC: Centre for Tourism Policy and Research, Simon Fraser University. Online: http://www.agf.gov.bc.ca/agritourism/publications/documents/agritourism_market_product_development_2001_fullrpt.pdf.

Willows, R.J., and R.K. Connell, eds. 2003. *Climate adaptation: Risk, uncertainty and decision-making*. Oxford: UK Climate Impacts Programme.

Wilson, A., and A. Tyrchniewicz. 1995. *Agriculture and sustainable development: Policy analysis on the Great Plains*. Winnipeg: International Institute for Sustainable Development (IISD).

Winkler J., J. Andresen, G. Guentchev, and R. Kriegel. 2002. Possible impacts of projected temperature change on commercial fruit production in the Great Lakes Region. *Journal of Great Lakes Research* 28 (4): 608-25.

Winkler, J., J. Andresen, G. Guentchev, J. Picardy, and E. Waller. 2000. Focus: Climate change and fruit production; An exercise in downscaling. In *Great Lakes regional assessment, preparing for a changing climate: The potential consequences of climate variability and change in the Great Lakes region (agriculture)*. http://www.geo.msu.edu/glra/assessment/assessment.html.

Wisner, B., P. Blaikie, T. Cannon, and I. Davis. 2004. *At risk: Natural hazards, people's vulnerability and disasters*. 2nd ed. New York: Routledge.

Wittrock, V., E. Wheaton, and C. Beaulieu. 2001. *Adaptability of Prairie cities: The role of climate. Current and future impacts and adaptation strategies*. SRC publication no. 11296-1E01. Saskatoon: Saskatchewan Research Council.

Wolfe, D. n.d. Potential impact of climate change on agriculture and food supply. Article available through United States Global Change Research Program, http://www.gcrio.org/USGCRP/sustain/wolfe.html.

World Bank. 2003. *Sustaining development in a dynamic world. Transforming institutions, growth, and quality of life*. Washington, DC: World Bank and Oxford University Press.

Yohe, G., K. Strzepek, T. Pau, and C. Yohe. 2003. Assessing vulnerability in the context of changing socioeconomic conditions: A study of Egypt. In *Climate change, adaptive capacity and development*, ed. J. Smith, R. Klein, and S. Huq, 101-36. London: Imperial College Press.

Zavaleta, E., M.R. Shaw, N.R. Chiariello, H.A. Mooney, and C.B. Field. 2003. Additive effects of simulated climate changes, elevated CO_2, and nitrogen deposition on grassland diversity. *Proceedings of the National Academy of Sciences* 100 (13): 7650-54.

Zentner, R., D. Wall, C. Nagy, E. Smith, D. Young, C. Miller, C.B. Campbell, B. McConkey, S. Brandt, G. Lafond, A. Johnston, and D. Derkson. 2002. Economics of crop diversification and soil tillage opportunities in the Canadian Prairies. *Agronomy Journal* 93: 216-30.

Ziska, L. 2004. Invasive weeds. In *Proceedings: Impacts of climate change on horticulture symposium*, 2 November 2003, Providence, RI. http://www.hort.cornell.edu/department/faculty/wolfe/cchortproc.pdf

Contributors

Younes Alila is Associate Professor, Department of Forest Resources Management, University of British Columbia, Vancouver, British Columbia.

Pierre André is Associate Professor, Département de Géographie, Université de Montréal, Québec.

Mark Barton is a climatologist with Environment Canada, Vancouver, British Columbia.

Suzanne Belliveau is a research associate in the Department of Geography, University of Guelph, Ontario.

Andrew Bootsma is Honorary Research Associate, Agriculture and Agri-Food Canada, Eastern Cereal and Oilseed Research Centre, Ottawa, Ontario.

Michael Brklacich is Professor, Department of Geography and Environmental Studies, Carleton University, Ottawa, Ontario.

Ben Bradshaw is Associate Professor, Department of Geography, University of Guelph, Ontario.

Chris Bryant is Professor, Département de Géographie, Université de Montréal, Québec.

Stewart Cohen is Senior Researcher, Adaptation and Impacts Research Division (AIRD), Environment Canada, Vancouver, British Columbia.

Wayne Caldwell is Associate Professor in Rural Planning in the School of Environmental Design and Rural Development, University of Guelph, Ontario.

Harry P. Diaz is Professor, Department of Sociology and Social Studies, University of Regina, Saskatchewan.

Grace Frank is a research assistant with Agriculture and Agri-Food Canada/PARC (Pacific Agri-Food Research Centre), Summerland, British Columbia.

Samuel Gameda is a research scientist with Agriculture and Agri-Food Canada, Ottawa, Ontario.

David Gauthier is Professor, Department of Geography, and Associate Vice President Research, University of Regina, Saskatchewan.

Walter Koch is a technical analyst with Source Water Protection, North Bay–Mattawa Conservation Authority, North Bay, Ontario.

Daniel McKenney is Chief, Geo-spatial Tools and Economic Analysis, Natural Resources Canada, Canadian Forest Service, Sault Ste. Marie, Ontario.

Robert McLeman is Assistant Professor, Department of Geography, University of Ottawa, Ontario.

Wendy Merritt is a research fellow with the Integrated Catchment Assessment and Management Centre, Australian National University, Canberra, Australia.

Denise Neilsen is a research scientist with Agriculture and Agri-Food Canada/ PARC (Pacific Agri-Food Research Centre), Summerland, British Columbia.

Cynthia Neudoerffer is a doctoral candidate in the Rural Studies PhD Program, University of Guelph, Ontario.

Susanna Reid is a planner with the County of Huron, Goderich, Ontario.

David Sauchyn is Research Professor, Prairie Adaptation Research Collaborative, University of Regina, Saskatchewan.

Bhawan Singh is Professor, Département de Géographie, Université de Montréal, Québec.

Barry Smit is Professor and Canada Research Chair, Department of Geography, University of Guelph, Ontario.

Scott Smith is a soil scientist with Agriculture and Agri-Food Canada/PARC (Pacific Agri-Food Research Centre) Summerland, British Columbia.

Bill Taylor is a climatologist with Environment Canada, Vancouver, British Columbia.

Henry D. Venema is Director, Sustainable Natural Resources Development, International Institute for Sustainable Development, Winnipeg, Manitoba.

Ellen Wall is Co-ordinator/Manager, C-CIARN Agriculture, Environmental Sciences, University of Guelph, Ontario.

David Waltner-Toews is Professor, Ontario Veterinary College, University of Guelph, Ontario.

Johanna Wandel is a postdoctoral research associate in the Department of Geography, University of Guelph, Ontario.

Index

Printed and bound in Canada by Friesens
Set in Stone by Artegraphica Design Co. Ltd.
Copy editor: Francis Chow
Proofreader: Megan Brand
Indexer: Annette Lorek